Markus Klinsmann

The Effects of Internal Stress and Lithium Transport on Fracture in Storage Materials in Lithium-Ion Batteries

Schriftenreihe
des Instituts für Angewandte Materialien
Band 54

Karlsruher Institut für Technologie (KIT)
Institut für Angewandte Materialien (IAM)

Eine Übersicht aller bisher in dieser Schriftenreihe erschienenen Bände finden Sie am Ende des Buches.

The Effects of Internal Stress and Lithium Transport on Fracture in Storage Materials in Lithium-Ion Batteries

by
Markus Klinsmann

KIT Scientific Publishing

Dissertation, Karlsruher Institut für Technologie (KIT)
Fakultät für Maschinenbau
Tag der mündlichen Prüfung: 23. September 2015

Impressum

KIT Scientific Publishing

Karlsruher Institut für Technologie (KIT)
KIT Scientific Publishing
Straße am Forum 2
D-76131 Karlsruhe

KIT Scientific Publishing is a registered trademark of Karlsruhe Institute of Technology. Reprint using the book cover is not allowed.

www.ksp.kit.edu

Print on Demand 2016

ISSN 2192-9963
ISBN 978-3-7315-0455-9
DOI 10.5445/KSP/1000050956

The Effects of Internal Stress and Lithium Transport on Fracture in Storage Materials in Lithium-Ion Batteries

Zur Erlangung des akademischen Grades

Doktor der Ingenieurwissenschaften

von der Fakultät für Maschinenbau
Karlsruher Institut für Technologie (KIT)

genehmigte

Dissertation

von

Dipl.-Phys. Markus Klinsmann
geboren am 28.11.1985
in Stuttgart (Deutschland)

Tag der mündlichen Prüfung: 23.09.2015

Hauptreferent: Prof. Dr. Marc Kamlah
Korreferent: Prof. Dr. Thomas Böhlke
Korreferent: Prof. Robert M. McMeeking

Abstract

Fracture of storage particles is considered to be one of the major reasons for capacity fade and increasing power loss in many commercial lithium-ion batteries due to emergent effects such as the growth of a solid electrolyte interphase, loss of contact in conductive pathways or complete disintegration of the electrode. The appearance of fracture and cracks in the particles is commonly ascribed to mechanical stress, which evolves from inhomogeneous swelling and shrinkage of the material when lithium is inserted or extracted. Here, we tackle the problem of fracture in storage particles by merging a coupled model of mechanical stress and diffusion of Li-ions with a phase field description of an evolving crack. To numerically solve the resulting differential equations, a parallelized finite element method computer code using adaptive mesh refinement and time step control is developed.

The phase field description for crack growth and fracture is an attractive alternative to numerical methods based on discrete representations of cracks, since the phase field methodology avoids the numerically challenging monitoring of the discontinuities introduced by the crack. In particular, for the simulation of complex crack growth topologies and application to coupled systems, e.g. with thermal or electrical fields, the phase field method has shown promise. However, an accurate prediction of the crack growth initiation is mandatory for a reliable simulation of crack trajectories both in terms of load history and the path followed through the material. We therefore first investigate predictions of crack growth derived from the phase field method and compare them with established relations from fracture mechanics.

The novel approach of simulating crack propagation coupled to diffusion of Li using a phase field method for fracture allows us to simultaneously study the evolution of the lithium concentration together with the initiation and growth of a crack in an arbitrary geometry, in two and three dimensions, and without presuming a specific crack path. We investigate how the formation of cracks depends on the size of the particle and an initial crack, as well as the applied flux at the boundary. Crack growth is observed both during lithium insertion and extraction in the first, as well as in the second half cycle. We find that dynamic effects resulting from the inertia of the material have a strong impact on the resulting crack patterns and describe the interplay between lithium diffusion and mechanical stress in the presence of a crack.

Zusammenfassung

Das Brechen von Speicherpartikeln wird als einer der schwerwiegendsten Mechanismen für Kapazitäts- und Leistungsverlust in vielen kommerziellen Lithium-Ionen-Batterien betrachtet. Gründe hierfür sind damit einhergehende Effekte, wie das Wachstum einer sogenannten *solid electrolyte interphase*, der Kontaktverlust an eine elektrisch leitende Anbindung oder die vollständige Zerrüttung der Elektrode. Die Entstehung von Rissen und Brüchen in den Partikeln wird üblicherweise mechanischen Spannungen zugeschrieben, die auftreten wenn sich das Material während des Einbringens oder Entnehmens von Lithium inhomogen ausdehnt oder zusammenzieht. In dieser Arbeit behandeln wir das Problem von Rissen in Speicherpartikeln indem wir ein gekoppeltes Modell für mechanische Spannungen und Lithium-Diffusion mit einer Phasenfeldbeschreibung des wachsenden Risses verbinden. Um das resultierende Modell numerisch zu implementieren wird ein parallelisierter Finite Elemente Computercode entwickelt, der sowohl eine adaptive Netzverfeinerung, als auch eine adaptive Zeitschrittweitenkontrolle unterstützt.

Die Phasenfeldmethode zur Beschreibung von Risswachstum ist eine attraktive Alternative zu anderen numerischen Methoden, die den Riss als diskretes Objekt darstellen. Dies liegt daran, dass die Phasenfeldbeschreibung das numerisch anspruchsvolle Verfolgen der bei einer diskreten Betrachtungsweise miteinhergenden Unstetigkeiten vermeidet. Infolgedessen hat sich die Phasenfeldmethode im speziellen bei Simulationen mit komplexen Risstopologien und in gekoppelten Problemen als vielversprechend erwiesen. Eine genaue Vorhersage der Rissinitiierung ist jedoch zwingend notwendig um verlässliche Vorhersagen für den exakten Rissverlauf durch das Material in Abhängigkeit von der angelegten Last zu treffen. Aus diesem Grund untersuchen wir im ersten Schritt das mittels der Phasenfeldmethode vorhergesagte Risswachstum und vergleichen es mit bewährten Beziehungen aus der Bruchmechanik.

Die neuartige Herangehensweise, Lithium-Diffusion und Risswachstum mittels eines Phasenfeldansatzes zu simulieren, erlaubt es gleichzeitig die Entwicklung der Lithium-Konzentration und die Initiierung und das Wachstum eines Risses zu studieren. Dabei liegt keine Einschränkung bezüglich der genauen Geometrie des Partikels oder der Anzahl an Raumdimensionen vor. Darüber hinaus sind keine Vorannahmen über den Verlauf des Risswachstums nötig. Wir untersuchen wie die Entwicklung eines Risses von seiner anfänglichen Länge und der Größe des Partikels, sowie dem aufgebrachten Fluss an der Oberfläche abhängt. Risswachstum wird sowohl während Zufuhr, als auch Entnahme von Lithium beobachtet. Dabei untersuchen wir neben dem ersten Halbzyklus auch die Frage, wie sich ein Riss im zweiten Halbzyklus bei umgedrehtem Oberflächenfluss entwickelt. Wir beobachten, dass die mit der Trägheit des Materials einhergehenden dynamischen Effekte eine starke Auswirkung auf die finale Rissgeometrie haben und beschreiben das Wechselspiel zwischen Lithium-Diffusion und mechanischen Spannungen in der Gegenwart eines teilweise stationären, teilweise wachsenden Risses.

Acknowledgements

The present thesis was written at the Institute for Applied Materials - Materials and Biomechanics (IAM-WBM) of the Karlsruhe Institute of Technology (KIT), with financial support from the Robert Bosch GmbH. Many people supported me during my studies and I would like to use these lines to express my gratitude to them.

My very special thanks go to my supervisors, Prof. Dr. Marc Kamlah and Prof. Robert McMeeking. I am deeply grateful for your support. The inspiring and encouraging discussions with you and the many valuable suggestions were more than helpful in achieving this thesis. In addition I would like to thank Prof. Robert McMeeking for having me as his guest for four months at the University of California Santa Barbara (UCSB). You went out of your way to make me have a wonderful time during my stay. The visit was financially supported by a scholarship of the Karlsruhe House of Young Scientists (KHYS).

Further, my sincere thanks go to Prof. Dr. Böhlke for agreeing to be the second referee of my thesis committee.

I express my heartfelt gratitude to Daniele Rosato, who supported my work within the Robert Bosch GmbH. I thank you for the many fruitful discussions we had, for always helping me to make good progress, for the proofreading and for your precious advices. It was a pleasure to have you as my adviser. I also thank Chiara Mastrodicasa for the great work and assistance during her internship at the Robert Bosch GmbH. Furthermore, I want to thank my colleagues at CR/ARM for the good collaboration during the past three years. The great atmosphere in the office made working on my thesis a very pleasant time.

For contributing to an unforgettable time during my stay in the USA, I would also like to thank Rob and Nancy Magee, as well as my Italian friends at UCSB and Garret Huff, who, in addition, assisted me with parts of the proofreading.

Above all, I deeply thank my dear friends and in particular my family for their support. This work could not have been achieved without having you on my side.

Contents

1 Introduction

1.1 Basics of Lithium-Ion Batteries

An electric battery consists of one or more electrochemical cells, which transform chemical energy into electrical energy [1, 2]. The electrochemical process involved in the conversion hereby determines the particular type of the battery. An irreversible process allows the battery to discharge only once, converting all the usable chemical energy into electrical energy. In contrast to such a primary battery, a secondary or rechargeable battery functions by a reversible process. Thus, the battery can be operated also in the opposite direction to charge it. By charging it, electrical energy is converted back to chemical energy and the battery is returned to a state, where it may be discharged again. We will specify the term 'process' in the following pages. Before doing so, we want to point out, that by denoting a process as irreversible, we are not only referring to the principal chemical reactions, which make up the working principle of a battery, but also to side reactions or other effects that lead to an overall irreversibility.

A second distinction considers the fundamental reactants of the electrochemical process. In today's most prominent battery type, the lithium-ion (Li-ion) battery, ionic lithium is participating in the relevant reactions. This is in contrast to Li-metal batteries, such as Li-sulfur or Li-air batteries, in which elemental lithium metal is formed. Other material systems are found in sodium-ion or nickel-metal hydride batteries.

Throughout this work, we will focus on the class of secondary Li-ion batteries. While a large variety of material systems is covered by this class, we will pose further restrictions with respect to certain material properties later in the text.

1.1.1 Fields of Application

The Li-ion technology has contributed a more than significant part to modern culture's mobile lifestyle. A convincing demonstration of this statement can be found in nearly everybody's pant pocket or handbag, the mobile phone. Taking the United States of America as example, with 326 million mobile subscribers in the year 2012 [3], on average each of the 314 million citizens was using wireless services, mainly via mobile phone. Since not only mobile phones, but also over 99% of tablets, notebooks and other mobile devices are powered by Li-ion batteries [4], practically everyone benefits from applications enabled by the Li-ion technology.

While consumer electronics still hold the biggest share of products equipped with Li-ion batteries, an increasing percentage of devices with Li-ion technology might also find their way into the hands of people, that by now were able to withstand the boon and bane of mobile communication. Examples are the rising usage of Li-ion chemistry in power tools, which should reach around 45% in 2015 [4] or the current growth of popularity in Li-ion powered electric bicycles [5], where German sales numbers nearly doubled from 200 000 in 2010 to 380 000 in 2012 [6].

Aside from that, strong growth is also being projected in the area of electric storage systems, where forecasts expect up to 2.6 GW of newly installed Li-ion power in the year 2020. With a total of worldwide 10.9 GW newly installed power, this would mean a share of around one fourth in a market, where competitors are not only alternative cell materials, but also completely different physical methods, such as for example 'Power to Gas', air compression, redox flow batteries or thermal energy storage. High expectations lie on these investments to promote and support the transition towards renewable energies and hopes can only feel encouraged when comparing the predicted numbers to the value of 0.1 GW that were installed in 2011 [7].

However, the greatest boost concerning the utilization of Li-ion technology is awaited in the automobile market, where Li-ion batteries provide the energy for electric propulsion systems. Depending on whether the electric powertrain is working solely or in combination with a conventional internal combustion engine, one differentiates between electric vehicles (EV) and hybrid (HEV) or plug-in hybrid electric vehicles (PHEV), that can also be charged by connection to an external power source. Forecasts in 2011 predicted a surge from around 14 % (3 639 GWh) to 37 % (49 034 GWh) of the total Li-ion energy storage demand for automobile applications in 2020, with a largest contribution by PHEVs (see Fig. 1.1). Yet, trends related to e-mobility are harder to predict than, for example, the development in the consumer electronics segment. This uncertainty is also caused by the strong impact of partly political factors, such as oil price development, infrastructure build-up for recharging possibilities or stricter emission standards for conventional combustion engine vehicles. Apart from that, new mobility concepts like car sharing in densely populated urban areas or special arrangements for commuters may have a non-negligible impact on the transition towards electrically propelled vehicles [8].

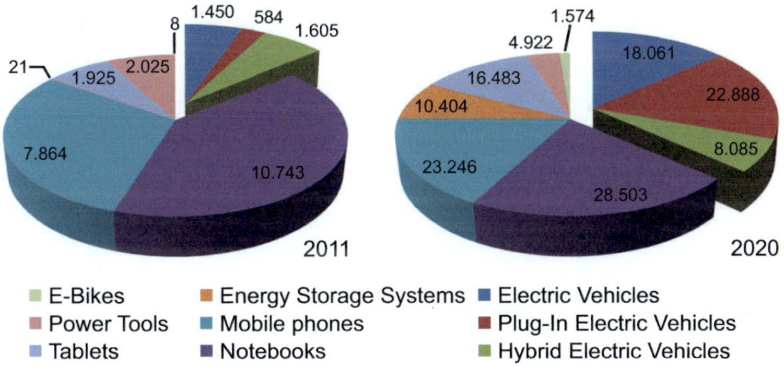

Figure 1.1.: Energy storage demand by volume and application in GWh. Extruded parts mark automotive applications. The values are extracted from diagrams in [7].

Despite all the imponderability, the key factor on the success of electrically powered automobiles and therefore also on the penetration of Li-ion technology in the automobile market is the Li-ion battery itself. Although first models are already being distributed quite successfully, there are still considerable challenges in order to convince the broad majority of saying goodbye to the combustion engine. One of the most obvious downsides that a potential customer faces when buying one of the currently available electric vehicles is undoubtedly the cost of the built-in battery. With costs of around 400 €/kWh [9] and a typical energy storage capacity of up to 20 kWh for a minicar [10, 11], not only a significant part of the costs for the electric powertrain, but also of the total acquisition costs of an electric vehicle is spent for the battery.

However, due to scale effects and improvements of the manufacturing productivity, batteries for automobiles should become cheaper in the coming years and prices below 120 €/kWh are awaited by 2025 [9].

More fundamental hurdles concern the battery technology. Better cells with increased energy density, both per volume and weight, have to be developed to improve the cruising range to levels where more extended journeys become feasible. For hybrid vehicles, where integration space is shared with the conventional propulsion system, higher power densities, mainly per volume, have to be obtained. For the usage outside of cities, fast-charge capability, enabling reasonable waiting times for intermediate recharge stops, is a further key ingredient. Due to the properties of certain chemical components, some Li-ion cells are susceptible to fires and explosions if the cell is damaged or suffers from production faults [12]. Thus, it is mandatory to assure safety for the passenger even in unusual situations. Aside from that, requirements concerning the lifetime of the battery are much more demanding in the automobile sector compared to, for example, consumer electronics. In the United States of America of 2010, a typical replacement cycle of a mobile phone lasted 22 months. It is not uncommon that the user already experiences a serious drop in the capacity of the built-in battery at the end of this period. For automobiles, it is desired that the costumer does not notice a significant performance loss over its full lifetime, i.e. for at least 10 years [13]. Since testing of cells over such a long duration is not practical, it is vital to establish a thorough understanding of the processes in a battery that lead to its degradation. Only then, reliable predictions can be made where real-time experiments over the full life cycle are not at disposition.

The described conditions for a wide success of Li-ion powered automobiles seem quite challenging at first glance. However, one has to keep in mind, that Li-ion is a comparatively young technology. While the history of compulsion engines in automobiles goes back to the end of the 19th century, batteries with Li were first proposed in the 1970s by Whittingham [14] and Besenhard et al. [15–17]. Almost two decades of pioneering works passed until the fist commercial Li-ion batteries were released by Sony in 1991 and a joint venture of Asahi Kasei and Toshiba in 1992 [18]. After two more decades, Li-ion technology is now on the verge of a breakthrough in the automobile market and the progress, which has been made during these years, can only strengthen the confidence that the last hurdles towards a completely electrified mobility may be overcome with Li-ion or one of its related technologies.

1.1.2 Design and Working Principle

The fundamental working principle of an electrochemical cell is the disparity in the electrochemical potential of charge carriers in different chemical compounds [1]. In a typical Li-ion battery, these compounds are, for example, on the one side lithium graphite Li_xC_6 and on the other side a lithium metal oxide, for example lithium manganese oxide $Li_{1-x}Mn_2O_4$, with x ranging from 0 to 1 [19]. If no external voltage is applied, the electrochemical potential of lithium is lower in the manganese oxide than in graphite and lithium has the higher tendency to form a compound with the manganese oxide. The reason for this is explained by a thermodynamic argument. At constant pressure and temperature, the change in Gibbs free energy is equal to the change of number of species multiplied by their chemical potential. In differential notation, this reads

$$\mathrm{d}G_{|p,T} = \sum_i \mu_i \,\mathrm{d}N_i, \tag{1.1}$$

with the Gibbs free energy G and the chemical potential and number μ_i and N_i of species i. In the presence of an electric field, an electric potential is added to the chemical potential to yield the electrochemical potential

$$\bar{\mu}_i = \mu_i + z_i F \Phi, \tag{1.2}$$

where z_i is the valency of the species, F is Faraday's constant and Φ is the local electrostatic potential. The second law of thermodynamics now states, that a spontaneous reaction is always following the direction, which minimizes the Gibbs free energy. The number of species with higher electrochemical potential thus decreases and the one of the species with lower electrochemical potential increases. This direction corresponds to the discharge process. By means of equation (1.2) we can reverse the direction through application of the appropriate voltage and charge the battery. For the above mentioned material combination this gives the reaction equation

$$\mathrm{Li}_x\mathrm{C}_6 + \mathrm{Li}_{1-x}\mathrm{Mn}_2\mathrm{O}_4 \underset{\text{charge}}{\overset{\text{discharge}}{\rightleftarrows}} \mathrm{C}_6 + \mathrm{LiMn}_2\mathrm{O}_4, \tag{1.3}$$

which consists of two steps. In the discharge (charge) direction, lithium is oxidized (reduced) at the graphite and reduced (oxidized) at the metal oxide

$$\mathrm{Li}_x\mathrm{C}_6 \underset{\text{charge}}{\overset{\text{discharge}}{\rightleftarrows}} \mathrm{Li}_x + \mathrm{C}_6 + x\,\mathrm{e}^-, \tag{1.4}$$

$$\mathrm{Li}_x + \mathrm{Mn_2O_4} + x\,\mathrm{e}^- \underset{\text{charge}}{\overset{\text{discharge}}{\rightleftarrows}} \mathrm{Li}_x\mathrm{Mn_2O_4}. \tag{1.5}$$

If these partial reactions take place at two spatially separated electrodes, both Li-ions and electrons have to move from one reaction site to the other. The electrode of the oxidation reaction is called anode, the one of the reduction reaction is labeled cathode. Although usually referred to the discharge of the battery, the denotation of cathode and anode can be misleading since it depends on the direction of the process. A stricter distinction is made on the basis of the reduction potential of the underlying reaction. The partial reaction with graphite (1.4) has a lower reduction potential than that with the manganese oxide (1.5). The former is therefore called negative electrode, the latter positive electrode.

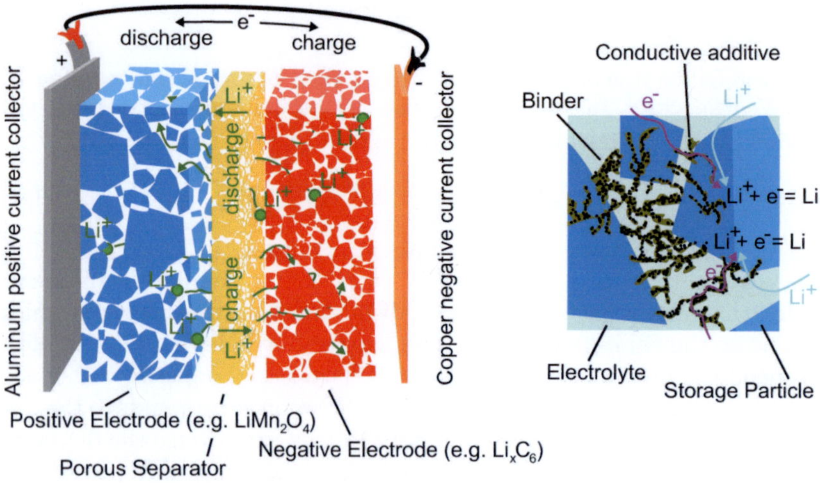

Figure 1.2.: (Left) Schmematic structure of an electrochemical cell design, consisting of current collectors, positive and negative electrode and separator. (Right) Illustration of the Li reduction at the surface of the storage particles. The Li-ions are transported through the electrolyte, which fills the pores in the separator and electrodes. Electron conduction takes place in the conductive network, consisting of conductive additives in the binder and storage particles.

In addition to the spatial separation of the electrodes, the second basic design element of an electrochemical cell is the enforcement of different pathways for the relevant charge carriers. If the electrons are not allowed to travel directly from one electrode to the other, but only through an external circuit, an electric current develops, which can be used to power a consumer [1]. For this purpose, an ordinary Li-ion battery consists of several components, as depicted in FIG. 1.2. A porous separator prevents direct electrical conduction between the electrodes and is most often made of polymers, such as polypropylene (PP) or polyethylene (PE) [2]. In order to provide ionic conduction, the pores of the separator are filled with a liquid electrolyte. Due to the high reactivity of lithium, they are based on non-aqueous, aprotic solvents. Examples are ionic liquids or organic compounds, such as propylene carbonate (PC) or dimethoxyethane (DME), with lithium hexafluorophosphate ($LiPF_6$) as conducting salt [2]. To establish electric conduction, the electrodes are in contact with metal foils made of copper and aluminum. These foils are called current collectors and serve as terminals to connect a consumer or voltage source to the cell.

The power, that a consumer can tap from a battery, and the time, that it takes to recharge it, depend in general only on one single condition. This is the availability of reactants for the chemical reactions in equation (1.4) and (1.5). Taking the discharge process at the cathode as an example, the condition can be stated in the following questions. How much Li-ions and electrons reach the reaction surface in a given time? How fast can Li-ions move into the material to allow more Li to flow through the surface and how large is the reaction surface itself? To optimize a battery with respect to these questions, a typical Li-ion battery electrode is built up as a composite of different materials, as shown on the right of FIG. 1.2. The reacting material is formed as so called active or storage particles with sizes below around $20\,\mu m$. These particles can be mono- or polycrystalline and effectively store the Li-ions [20, 21]. The storage particles are embedded in a soft, porous binder material, which provides structural integrity of the electrode and allows electrolyte to reach the particle surfaces through its pores. In most cases, it consists of a polymeric material, such as polyvinylidene difluoride (PVDF). But also alternative concepts, such as alginates or rubber materials, are being pursued [1, 22]. To ensure a high electronic conductivity of the electrode, the binder is enriched by so called carbon black (a nano-sized carbon additive) [2]. Particles and carbon black build an electrically conductive network through which the storage particles are connected to the current collector.

The choice of active material determines the specific type of Li-ion cell. The above mentioned graphite and manganese oxide compounds belong to the class of intercalation materials. Members of this class have in common, that the structure of the material remains unchanged during insertion of the lithium. The material thus acts as a host providing empty lattice sites that can be occupied by the lithium. This is in contrast to, for example, alloying materials [23], such as silicon or tin anodes, or conversion materials, such as copper fluoride CuF_2 [24], which lead to completely new crystal structures or chemical compounds. An alternative example for an intercalation material on the anode side is lithium titanate $Li_{4+x}Ti_5O_{12}$. It is structurally more stable and more benign to reactions with the electrolyte than graphite [25, 26], but possesses a higher electrochemical potential leading to a lower overall cell voltage. On the cathode side, different metal oxides, like the aforementioned lithium manganese oxide $LiMn_2O_4$, lithium cobalt oxide $LiCoO_2$ or lithium nickel oxide $LiNiO_2$, all come with their own advantages and disadvantages [20, 21]. They are therefore often used in combination as so called NCM ($LiNi_xCo_yMn_zO_2$) or as blend with particles of different materials [27]. A further intercalation material, which does not fall under the class of metal oxides, is lithium iron phosphate $LiFePO_4$. It is thermally more stable, but comes with a lower electrochemical potential than the metal oxides [21].

1.1.3 Aging and Degradation Mechanisms

The primary tasks of both building durable lithium (Li) ion batteries, that preserve capacity over a large number of cycles, and predicting the remaining life time of a commercial Li-ion battery demands a thorough understanding of the underlying mechanisms leading to the degradation of a battery. As expected for a battery, a large part of these degradation mechanisms are of electrochemical nature [28].

Among these detrimental effects, one of the most severe contributions is ascribed to the formation of the so called solid electrolyte interphase (SEI) [29–36]. In Li-ion batteries, the SEI is a layer that forms on the surface of the active particles in side reactions with the electrolyte, mainly during the first few cycles. Its composition is highly dependent on the choice of electrolyte, additives and active material. However, since Li is always one of its constituents, the build-up of the SEI reduces in any case the amount of available Li for energy storage.

Furthermore, it provides an obstacle for Li to reach the surface of the storage particles, which results in an increased electrode impedance. It is thus a crucial task to achieve a SEI with a low irreversible capacity loss, that is stable in the sense, that its growth saturates at a certain thickness and that does not detach from the particle surface [28].

Another chemical degradation mechanism leading to capacity fade is the dissolution of active material, which presents a severe problem for example in electrodes using lithium manganese oxide (LMO) as active material [29, 37–39]. It is commonly assumed that the manganese undergoes a disproportionation reaction in which trivalent Mn(III) becomes di- and tetravalent Mn(II) and Mn(IV). The Mn(II) dissolves into the electrolyte [37] and may then participate in or cause further SEI reactions [40]. The process thus presents a cause for both loss of active material and raise of the electrode impedance.

The mentioned disproportion reaction can be considered as one example of an undesired phase transition in the active material. A further example of an unwanted phase change appears in metal oxides with special layered structures to obtain higher operating voltages [41–44]. Due to the higher voltages, these materials are identified as promising candidates to improve the energy density of Li-ion batteries. However, at the current stage of development, the materials are prone to drastic voltage drops because of structural rearrangements already in the first cycles.

Another aging effect is the deposition of elemental Li on the electrodes in form of plating or growth of dendrites [45, 46]. The phenomenon not only causes capacity fade, but may also lead to complete failure of the battery when dendrites cause an electrical short. This becomes most severe during over-charge or operation at very low temperatures [47, 48].

Further chemical degradation mechanisms are less related to the material of the storage particles. For example, the current collectors may corrode, electrolyte can decompose and lead to gas evolution or the binder may degrade [28].

Although already the chemical degradation mechanisms are numerous and complex to understand, in the recent past, a growing scientific interest reflects the rising awareness that also mechanical aspects influence lifetime behavior of a Li-ion battery in a more than non-negligible way. While some effort is spent on the impact of mechanical stress on the macroscopic scale of a Li-ion battery [49–53], the greater focus lies on the description of the interplay between diffusion and stress within electrode storage particles. Our own work is settled in this field and we will go into more details in the following section.

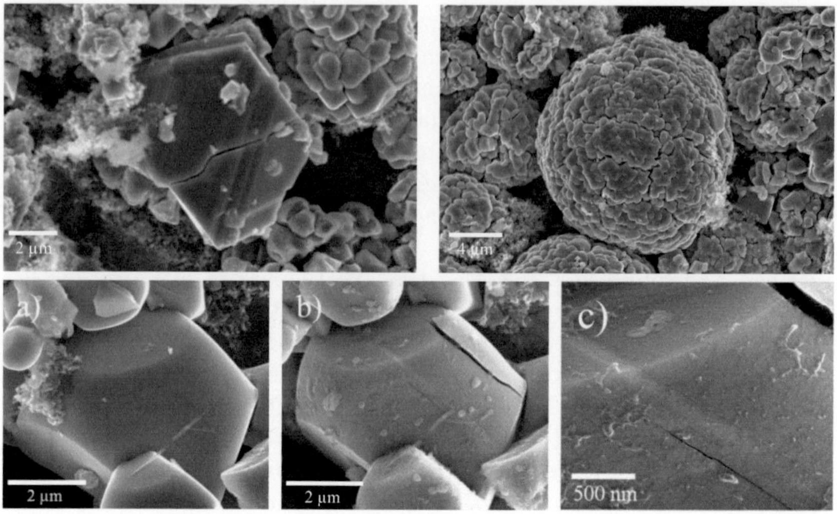

Figure 1.3.: Pictures taking via scanning electron microscopy. Top row: Crack in LMO (left) and fracture/delamination between primary particles in NCM (right) after cycling. Shown by kind permission of T. Waldmann, ZSW Baden-Württemberg. Bottom row: LMAO Pristine Particle ($LiMn_{1.95}Al_{0.05}O_4$) before and after 800 cycles [54].

1.2 State of the Art

The microscopic stress appearing in storage particles results from an inhomogeneous volume change, which the material encounters when Li is extracted from or inserted into the particle. Depending on the type of material, the volume change is isotropic as in $LiMn_2O_4$ [19] or anisotropic as in $LiCo_{1/3}Ni_{1/3}Mn_{1/3}O_2$ [55] and can exceed 300% as in Si [56]. As a grave outcome of such stress, fracture and complete breakage of storage particles may occur [57–60]. Examples for the observation of fracture in storage particles are shown in FIG. 1.3. Subsequent loss of electrical contact for the fractured particles [61, 62] and growth of the solid electrolyte interphase on the new surfaces [63] are blamed as fatal threats to a Li-ion battery expected to have a long life in terms of capacity fade and power loss [64, 65]. Detailed insight into the problem of fracture in storage particles thus presents a central key to improvements in battery design and appropriate operation for future generations.

One of the earliest efforts on modeling stress in active particles is the work of Christensen et al. [66]. Included in their description is the effect of mechanical pressure on Li diffusion, as well as nonideal interactions between Li and host material. Reduced in complexity and valid only for dilute Li concentrations, but applicable to fully three-dimensional particles, a model of Zhang et al. [67] followed, which may be regarded as the basis of many subsequent variations on their approach. Among these are analytic works deriving closed form relationships among relevant quantities by either geometric or physical simplification [68, 69] or mathematical linearization [70]. Solutions of this kind are also found for core-shell like particles [71], a promising approach for materials with very large volume expansion, in which critical stress levels may be avoided by using a composite structure. Corresponding numerical investigations are presented in [72, 73]. The core-shell composite structure should not be confused with phenomena in homogeneous materials that may experience a core-shell like concentration distribution during Li insertion or extraction [74]. A general strategy for avoiding high mechanical stress caused by lithiation swelling and shrinkage is to use smaller particles sizes, possibly nanometer scale [75–77]. At such small sizes, surface effects become more dominant [78–82] and can even lead to the suppression of phase separations [83–85]. Phase changes are also modeled by means of phase field models of Cahn-Hilliard type [86–88] and demonstrate that large stress may also result at low C-rates.

On the other hand, more elaborate mechanical treatments have been undertaken, such as those that included elastic-plastic material behavior or formulations for large deformations [89–93] as well as the introduction of dislocations [94]. Going one step further, some treatments take the computed stress and strain resulting from Li diffusion and use them to investigate the critical conditions under which a particle may fracture. For example, Woodford et al. [95] develop a so called electrochemical shock map, sketching the values of parameters for which single crystals and polycrystals tend to crack. A similar map is provided in [96] for a phase-transforming material. Woodford et al. [95, 96] support their findings by acoustic emission experiments, a method for monitoring particle fracture during cell operation [97–99] and a potential candidate for linking it with decreased battery performance [100]. A basic concept for determining the circumstances leading to cracks is a comparison of the stored elastic energy released from the bulk material and the energy required to create new surfaces [101]. The comparison is often evaluated theoretically by introduction of an artificial crack in the material and followed by computation of the energy required to grow the crack [102–105]. Studies of systems with preexisting cracks address, for example, the influence of the open circuit potential [106] or of plastic yielding [107] on fracture resistance.

In this work, we aim at simulating crack propagation coupled to diffusion of Li using a phase field method for fracture. Among the most complicated aspects of the theory of fracture is the prediction of the path to be taken by one or several cracks in a specimen or structural component. Due to the complexity of the phenomena involved, one commonly relies on computational methods to approach the problem [108]. A modeling approach that allows a comparatively convenient numerical implementation is the phase field method for fracture. It is identified as a promising candidate for studying intricate crack growth problems, including effects such as crack branching, kinking or merging in arbitrary geometries and dimensions [109, 110].

Similar to phase field methods used for other problems is the inherent existence of one or many order parameters that take defined values depending on the state and development of the phases subject to numerical smoothing, giving a continuous transition between states. In the area of fracture, the order parameter discriminates between an intact and a cracked region of the material. In contrast to discrete crack models, the phase field does not attempt to describe the crack as a geometric feature with a physical discontinuity, but as a smoothed continuum approximation with a gradient providing a transition between the intact and fractured states, taking place within a characteristic length, l. [111, 112].

The crucial advantage of the phase field approach is the avoidance of complicated, sophisticated techniques necessary to describe complex crack topologies in the discrete setting. Examples of such techniques are the adaptive reconstruction of the mesh through introduction of new boundaries at freshly created crack surfaces [113] and the extended finite element method (XFEM) [114], that enriches the shape functions in cracked elements by addition of a set of discontinuous ones. Also rather common is the usage of so called cohesive elements that allow displacement jumps at element boundaries and therefore restrict the crack to penetration along the corresponding element edges [115, 116].

Due to the smooth character of the phase field, complicated crack patterns, as for example shown in [117–119], are in principle achievable in the setting of a standard finite element method. The easy implementation of the method resulting from this benefit has lead to its utilization already beyond purely mechanical problems, such as for example with thermal expansion [120], in ferroelectrics [121, 122] or in piezoelectrics [123, 124]. The flexibility of the approach has even been exploited in a coupled model with water diffusion [125].

1.3 Objectives and Overview

In this work, we seek a deeper understanding of the phenomena involved in the comminution of storage particles of Li-ion batteries by coupling a model of Li diffusion and mechanical stress with a phase field approach to fracture. A similar combination is found in [126] for thin films and in [127] for $LiFePO_4$ nanoparticles. While [126] concentrates on different arrangements of cracks in a 2D setup, [127] investigates the influence of detailed material properties, such as, for example, anisotropic coherence strains and anisotropic Li diffusion. Here, we present an alternative formulation. It includes a thermodynamically consistent description of Li mass transport that is valid beyond the dilute case and takes into account mechanical stress as an additional driving force for diffusion. The formulation is combined with a fully dynamical description of the fracture process, and we find that dynamic effects resulting from the inertia of the material have a strong impact on the resulting crack patterns, and may even lead to complete breakage of a particle in one half cycle of insertion, or in only a few half cycles. The work is organized as follows.

In **chapter 2**, the physical model is derived. We first describe the interplay of Li diffusion and mechanical stress and then introduce the phase field method for crack growth. Thereafter the two parts of the model are coupled.

The resulting system of partial differential equations demands for a sophisticated solution scheme, which is presented in **chapter 3**. We determine the weak form of the boundary value problem and cast it into a form that can be implemented into a computer algorithm. We then describe the general solution scheme and develop a mesh adaptation scheme, that is both based on an energy criterion and the evaluation of the gradient of the phase field. In addition, a time-adaptive method is presented, that keeps track of the currently dominant physical time scale and adjusts the numerical time steps accordingly.

In **chapter 4**, the phase field method for crack growth is compared with established results from fracture mechanics. We examine the accuracy of predictions on crack initiation and propagation in a pure bending and an edge notch stretch test and investigate two different methods for the introduction of an initial crack. In addition, we study the role of the length scale parameter as a material parameter and the practicability of using it for the definition of an initial crack distribution in **chapter 5**.

Chapter 6 and **7** present a comprehensive study on Li diffusion and fracture in spherical and cylindrical particles. Precracks are introduced into the particles and the fracture behavior during both Li insertion and extraction is investigated. Several phenomena, such as crack branching or complete breakage of the particles, are found and the difference between two and three-dimensional setups are examined. Further, in **chapter 8** we have a look on how cracks, that have grown in the first half cycle, may extend in a second half cycle with reversed boundary conditions.

We give a quick outlook on how the model could be extended with respect to the Li flux at the crack in **chapter 9** and conclude our work in **chapter 10**.

2 Model Formulation

2.1 Stress and Diffusion in a Storage Particle

In this chapter a model is set up, which describes Li diffusion in an intercalation material. The basis of the model is found in the work of Bohn et al. [128, 129], which considers flux of Li in a phase-changing material. However, we neglect phase transformations and reformulate the model to make it more suitable for extension to include a fracture phase field order parameter.

The central entity of the model is a single free energy density potential from which we can derive all our material relations via partial derivatives. In order to arrive at the free energy density potential, the following assumptions are imposed:

- a fixed host lattice in the sense that as Li diffuses into the material the lattice remains unaltered (intercalation material)
- intercalating Li is neither direct nor indirect interacting with each other (ideal solution)
- taking into account the availability of lattice sites for hopping of Li and saturation effects (non-dilute solution)
- a constant and homogeneous temperature throughout the particle
- an isotropic and linear volume expansion due to the inserted Li
- an isotropic and linear elastic material behavior
- the strain in the material is small in the sense that the mechanics of infinitesimal straining provide a good approximation

We return to these assumptions and consider them in more detail after first deriving the fundamental balance equations, that hold independent of the specific material system.

2.1.1 Balance Equations

Balance of Li content

The time rate of change of the total content of Li stored in a part of a particle is equal to the amount flowing in or out of it. Herein lies the assumption that there do not exist mechanisms that stop Li in its diffusive motion, for example by being trapped to certain lattice sites. Further, no previously immobile Li is released for diffusion. Denoting the domain describing a part of the body $\mathcal{B}$ with $\mathcal{P} \subseteq \mathcal{B}$ and its boundary with $\partial\mathcal{P}$, this reads

$$\frac{\mathrm{d}}{\mathrm{d}t}\int\limits_{\mathcal{P}} \mathrm{d}V\, c(\mathbf{x},t) = -\int\limits_{\partial\mathcal{P}} \mathrm{d}S\, \mathbf{J}(\mathbf{x},t)\cdot\mathbf{n}(\mathbf{x},t). \tag{2.1}$$

Here, t is a scalar denoting the time, $\mathbf{x}$ a vector in space, $c(\mathbf{x},t)$ the Li concentration, $\mathbf{J}(\mathbf{x},t)$ the Li flux and $\mathbf{n}(\mathbf{x},t)$ a unit vector normal to the boundary of the domain. Working in the regime of small strains, we can neglect a change of the shape or size of the domain with time. We therefore take the time derivative[1] into the integral yielding

$$\int\limits_{\mathcal{P}} \mathrm{d}V\, \dot{c}(\mathbf{x},t) = -\int\limits_{\partial\mathcal{P}} \mathrm{d}S\, \mathbf{J}(\mathbf{x},t)\cdot\mathbf{n}(\mathbf{x}) = -\int\limits_{\mathcal{P}} \mathrm{d}V\, \nabla\cdot\mathbf{J}(\mathbf{x},t) \tag{2.2}$$

where we have used the divergence theorem with the nabla operator defined by $\nabla = (\partial/\partial_x, \partial/\partial_y, \partial/\partial_z)$ and the short notation of the time derivative given by $\dot{c}(\mathbf{x},t) = \frac{\mathrm{d}}{\mathrm{d}t}c(\mathbf{x},t)$. Since this relation holds for any arbitrarily small part of the body, it is equivalent to state the balance equation in differential form by

$$\dot{c}(\mathbf{x},t) + \nabla\cdot\mathbf{J}(\mathbf{x},t) = 0. \tag{2.3}$$

[1] In this case, the material time derivative is required. Since assuming small strains, material time derivatives are computed via partial derivative with respect to t.

Balance of Linear Momentum

The time rate of change of the linear momentum of a body (or a part of it) is equal to the sum of forces acting on it. We neglect body forces such as, for example, due to gravity, and consider only forces that are applied on the boundary of the body. The balance of linear momentum then requires

$$\frac{\mathrm{d}}{\mathrm{d}t}\mathbf{P} := \frac{\mathrm{d}}{\mathrm{d}t}\int_{\mathcal{P}} \mathrm{d}V\, \rho(\mathbf{x},t)\mathbf{v}(\mathbf{x},t) = \int_{\partial\mathcal{P}} \mathrm{d}S\, \mathbf{t}(\mathbf{x},t,\mathbf{n}) \tag{2.4}$$

Here, the total linear momentum $\mathbf{P}$ is defined as the integral of the mass density $\rho(\mathbf{x},t)$ times the velocity field $\mathbf{v}(\mathbf{x},t)$, which, in the small strain regime, is given by the partial time derivative of the displacement field $\mathbf{v}(\mathbf{x},t) = \dot{\mathbf{u}}(\mathbf{x},t)$. Since Li is a comparatively light element (e.g. in $LiMn_2O_4$, the atomic weights are 6.94 u for Li, 54.94 u for Mn and 16.00 u for O), the mass density is taken to be independent of the Li concentration. Further, in the small strain limit, elastic compression or expansion does not have a significant impact on the density. It is therefore taken as a constant and we can move the time derivative into the integral and apply it only to the velocity. For the right hand side, Cauchy's theorem states, that the traction vector $\mathbf{t}(\mathbf{x},t,\mathbf{n})$ can be written in terms of the Cauchy stress tensor as

$$\mathbf{t}(\mathbf{x},t,\mathbf{n}) = \boldsymbol{\sigma}(\mathbf{x},t)\mathbf{n}. \tag{2.5}$$

Applying the divergence theorem, we obtain

$$\int_{\mathcal{P}} \mathrm{d}V\, \rho\,\dot{\mathbf{v}}(\mathbf{x},t) = \int_{\mathcal{P}} \mathrm{d}V\, \nabla\cdot\boldsymbol{\sigma}(\mathbf{x},t). \tag{2.6}$$

As before, the mentioned arguments not only hold for the full body, but for any part of it. We may therefore express it in differential form by

$$\rho\,\dot{\mathbf{v}}(\mathbf{x},t) = \nabla\cdot\boldsymbol{\sigma}(\mathbf{x},t). \tag{2.7}$$

2.1.2 Free Energy Density

Equations (2.3) and (2.7) require constitutive relations, that describe the physical behavior of the material, relating the secondary variables

$\mathbf{J}$ and $\boldsymbol{\sigma}$ with the principle variables concentration c and displacement field $\mathbf{u}$. Following principles of the so called theory of materials [130] and confining ourselves to the case of infinitesimal strain, $\boldsymbol{\sigma}$ depends only on the symmetric part of the gradient of $\mathbf{u}$. Constitutive relations are therefore given in terms of the small strain tensor

$$\boldsymbol{\varepsilon} = \frac{1}{2}\left[\nabla\mathbf{u} + (\nabla\mathbf{u})^T\right]. \tag{2.8}$$

For a compact and easily extendible formulation of the model, we define a single free energy density potential ψ from which we obtain the constitutive relations through partial derivatives with respect to their conjugate variables by $\boldsymbol{\sigma} = \partial_{\boldsymbol{\varepsilon}}\psi$ for the stress and $\mu = \partial_c\psi$ for the chemical potential. The appropriate free energy function is derived from a number of considerations that are deduced from certain physical assumptions. In the following, these shall be explained step by step.

Mechanical Part of Free Energy Density

We commence with the elastic energy stored in the material, i.e. ψ_{elastic}, and use an analogy to a material undergoing thermal expansion proportional to the local temperature. For an isotropic linear elastic material, the elastic energy is then obtained by

$$\psi_{\text{elastic}}(\boldsymbol{\varepsilon}, c) = \frac{1}{2}L(\text{tr}[\boldsymbol{\varepsilon} - \boldsymbol{\varepsilon}_{\text{Li}}])^2 + G'\text{tr}[(\boldsymbol{\varepsilon} - \boldsymbol{\varepsilon}_{\text{Li}})^2], \tag{2.9}$$

with Lamé's first parameter L, the shear modulus G' and the trace operator $\text{tr}[\boldsymbol{A}] = \sum_i A_{ii}$. The strain caused by the isotropic volume expansion due to the intercalated Li is given by

$$\boldsymbol{\varepsilon}_{\text{Li}} = \frac{\Omega}{3}(c - c_{\text{ref}})\mathbf{1}, \tag{2.10}$$

where Ω is the molar volume of Li and c_{ref} the reference concentration.

Chemical Part of Free Energy Density

Next, we consider the change of the internal energy density, which occurs when Li is introduced into the host lattice. In contrast to the foundation of our model [128], we neglect mutual interaction between Li in this work and obtain

$$\psi_{\text{chem}}(c) = \mu_0 c, \tag{2.11}$$

where μ_0 is the reference chemical potential of Li in the host material, which is independent of pressure, temperature or concentration. Since this term only adds a constant value to the full chemical potential of Li, it does not enter the diffusion equation. However, one can expand ψ_{chem} by more complex terms to account for phenomena like phase changes.

Entropic Part of Free Energy Density

The last contribution to the free energy density stems from the entropy, resulting from the disorder of Li-ions in the host lattice. Using combinatorics, we count the number of possibilities $\mathcal{W}$ of arranging N Li-ions on $\hat{N}$ host lattice sites yielding the entropy density

$$\eta = \frac{1}{V_0} k_B \ln \mathcal{W} = \frac{1}{V_0} k_B \ln \frac{\hat{N}!}{N!(\hat{N}-N)!}, \tag{2.12}$$

where V_0 is a reference volume and k_B the Boltzmann constant. After approximation of the expression for large values of $\hat{N}$, we replace Li-ion numbers by concentrations and then integrate it to receive

$$\psi_{\text{entropic}}(c) = -R\theta \int_0^c \mathrm{d}c' \ln \frac{1 - c'/c_{\max}}{c'/c_{\max}}. \tag{2.13}$$

Here, R is the universal gas constant and θ denotes the temperature, which is assumed constant throughout the particle.

Full Free Energy Density

The expressions given in equation (2.9), (2.11) and (2.13) describe the different contributions to the free energy density that result from the assumptions presented in the introduction of this chapter. Combining these, we obtain the full free energy density

$$\begin{aligned}\psi(\varepsilon, c) = {} & \frac{1}{2} L (\mathrm{tr}[\varepsilon - \varepsilon_{\mathrm{Li}}])^2 + G' \mathrm{tr}[(\varepsilon - \varepsilon_{\mathrm{Li}})^2] \\ & + \mu_0 c - R\theta \int_0^c \mathrm{d}c' \ln \frac{1 - c'/c_{\max}}{c'/c_{\max}}.\end{aligned} \tag{2.14}$$

2.1.3 Constitutive Relations

With the free energy density given by (2.14), the constitutive relations are computed through partial derivatives, as described above, yielding

$$\boldsymbol{\sigma} = L\,\mathrm{tr}[\boldsymbol{\varepsilon} - \boldsymbol{\varepsilon}_{\mathrm{Li}}]\mathbf{1} + G'\,(\boldsymbol{\varepsilon} - \boldsymbol{\varepsilon}_{\mathrm{Li}}), \tag{2.15}$$

$$\mu = -\Omega\,\sigma_h + \mu_0 - R\theta \ln \frac{1 - c/c_{\mathrm{max}}}{c/c_{\mathrm{max}}}, \tag{2.16}$$

where we have used the hydrostatic stress given by

$$\sigma_h = \frac{1}{3}\mathrm{tr}[\boldsymbol{\sigma}]. \tag{2.17}$$

The so defined stress-strain relationship (2.15) allows to solve the equation describing conservation of linear momentum (2.7) for any given Li concentration. In contrast, the equation for balance of Li content (2.3) requires as a further ingredient a relationship between the chemical potential and the Li flux. This is generally given in the form [67]

$$\mathbf{J} = c\,\mathbf{v}_{\mathrm{Li}} = -c\,M(c)\nabla\mu, \tag{2.18}$$

where $\mathbf{v}_{\mathrm{Li}}$ is the velocity of Li atoms and M is their mobility. The expression can be understood as follows. The flux of Li is equal to the amount of Li per reference volume, i.e. the concentration c, flowing with a certain velocity $\mathbf{v}_{\mathrm{Li}}$. The driving force for the development of the velocity field is given through the gradient of the chemical potential μ. Depending on the mobility, the Li motion follows this driving force more or less rapidly. A common approach is to take the mobility to be constant [67, 131]. However, given Li hopping in a host lattice, it is reasonable to assume that the mobility of Li should decrease with increasing lattice site occupancy. In the most simple approximation, this implies a linear relation of the form

$$M(c) = M_0(1 - c/c_{\mathrm{max}}), \tag{2.19}$$

where $M_0 = D_0/(R\theta)$ is the mobility for a dilute solution of Li in the host material and D_0 denotes the diffusion coefficient of Li. Combining equations (2.16), (2.18) and (2.19) we then obtain the flux

$$\mathbf{J} = -D_0\nabla c + \frac{D_0\Omega}{R\theta} c\,(1 - c/c_{\mathrm{max}})\nabla\,\sigma_h. \tag{2.20}$$

The first term on the right is consistent with Fickian diffusion that would result from constant mobility and concentrations of Li smaller than c_{max}. However, given concentration dependent mobility as stated in equation (2.19) it prevails at any concentration of Li. In addition, the concentration dependent mobility of Li as given in equation (2.19) ensures the significance of stress gradient driven diffusion controlled by the second term on the right hand side of equation (2.20) and determines that it is negligible at low and high Li concentration. As can be deduced from equation (2.20) a constant mobility would lead to an overestimation of the stress gradient driving force at higher Li concentration.

We additionally need a set of initial and boundary conditions, defined on the domain B describing the storage particle geometry and its boundary ∂B. We denote prescribed displacements by $\bar{\mathbf{u}}$, tractions by $\bar{\mathbf{t}}$, concentrations by $\bar{c}$ and fluxes by $\bar{\mathbf{J}}$. A well-defined set of boundary conditions is given by $\mathbf{u} = \bar{\mathbf{u}}$ on ∂B_u and $\mathbf{t} = \boldsymbol{\sigma n} = \bar{\mathbf{t}}$ on ∂B_t with $\partial B_u \cap \partial B_t = \emptyset$ and $\partial B_u \cup \partial B_t = \partial B$ for the conservation of linear momentum, as well as $c = \bar{c}$ on ∂B_c and $\mathbf{J} = \bar{\mathbf{J}}$ on ∂B_J with $\partial B_c \cap \partial B_J = \emptyset$ and $\partial B_c \cup \partial B_J = \partial B$ for the conservation of Li. The required initial conditions have the form $c(\mathbf{x}, t_0) = c_0(\mathbf{x})$, $\mathbf{u}(\mathbf{x}, t_0) = \mathbf{u}_0(\mathbf{x})$ and $\mathbf{v}(\mathbf{x}, t_0) = \mathbf{v}_0(\mathbf{x})$.

2.2 Fracture in a Phase Field Approach

The classical energy criterion for brittle fracture developed by Griffith and Irwin [132–134] states that a preexisting crack starts to grow as soon as the elastic energy in the material that can be released during crack propagation is sufficient to overcome the fracture resistance of the material. That is, to extend a crack of length a by a distance Δa, enough stored energy must be transformable into surface energy, plastic work and other dissipative processes required to separate the material. From a thermodynamical point of view, crack extension thus can be treated as a process that occurs, subject to irreversibility, to minimize a free energy composed of the stored elastic bulk energy plus the fracture energy. The foundation of the fracture phase field method through such a variational approach was established in [135] and [136]. Approximate representations of the energy variation were then found by introduction of a scalar field d and a length scale parameter l transverse to the crack surface, so that the original surface based fracture energy functional was characterized as a volume based parameter over the distance l in a regularized form in terms of d and its gradient [137], an idea originally developed for problems related to image segmentation [138]. A requirement is that the original energy functional should be restored in the sense of Γ-convergence when l tends to zero. A comprehensive overview of the topic is given in [139].

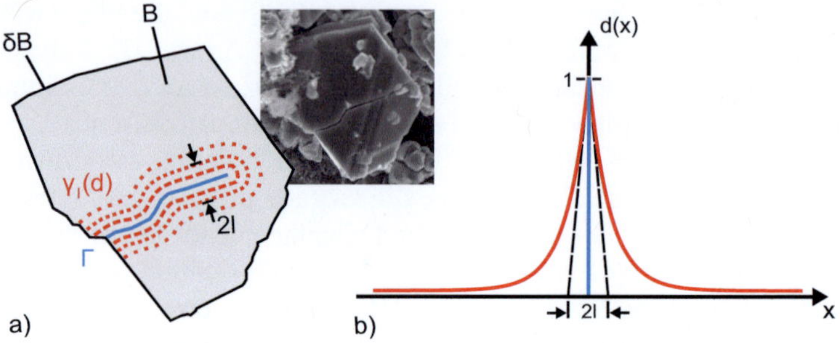

Figure 2.1.: (a) Schematic illustration of the characterization of a cracked region by $d = 1$ along Γ, in blue, and smoothing of the uncracked-cracked transition over a length scale l, in red. (b) Corresponding one dimensional representation. A scanning electron micrograph of a $LiMn_2O_4$ particle is also shown (permission of T. Waldmann, ZSW Baden-Württemberg).

Here, we follow the formulation of rate-independent crack propagation given in [140], with modifications taken from [117]. Alternative approaches are found, for example, in [112, 141].

A standard feature of a phase field method is the definition of a so called order parameter that indicates the state of a phase present at a certain point in time $t \in \mathbb{R}$ and space $\mathbf{x} \in \mathbb{R}^{\text{dim}}$. In our case, the two states refer to cracked and intact material and correspond to order parameter values 1 and 0 respectively, as depicted for a discrete crack by the blue line in FIG. 2.1. Thus, the order parameter[2] is of the form

$$d(\mathbf{x}) = \begin{cases} 1, & \text{if material is cracked,} \\ 0, & \text{if material is intact.} \end{cases} \tag{2.21}$$

The second distinctive feature of the method is a spatially smoothed representation of the crack, given that x is distance measured orthogo-

[2] We will use both the term order parameter and phase field to denote $d(\mathbf{x}, t)$ without any distinction.

nally from the center of the smeared crack. A typical continuous, smooth approximation of (2.21) in one dimension is the exponential function

$$d(x) = e^{-|x|/l}, \tag{2.22}$$

with the length scale parameter $l \in \mathbb{R}^+$ determining the width of smearing of the function and thus the distance over which material transitions smoothly from the uncracked to the cracked state. As l approaches zero, a discontinuous function is recovered representing a discrete crack. The smoothed approximation to the spike function is depicted in FIG. 2.1 b). It can be shown that the function in equation (2.22) is a minimizer of the one-dimensional analog to the functional

$$W_{\text{crack}}(d) = G_c \int_{\mathcal{B}} \mathrm{d}V \underbrace{\left(\frac{1}{2l} d^2 + \frac{l}{2} |\nabla d|^2 \right)}_{\gamma_l(d)}, \tag{2.23}$$

subject to the condition that there is a crack at $x = 0$. Here, G_c is the Griffith-type critical energy release rate and $\gamma_l(d)$ is the so called crack surface density [140]. A solution to the minimization principle

$$d(\mathbf{x}, t) = \text{Arg} \left\{ \inf_{d \in D_{\Gamma(t)}} W_{\text{crack}}(d) \right\}, \tag{2.24}$$

thus represents a smoothed approximation to the discrete crack at $\mathbf{x} \in \Gamma(t)$, as depicted in FIG. 2.1 a). The functional W_{crack} approximates the fracture energy through substitution of the integration along the crack surface $\Gamma(t)$ by a volume integration of the crack surface density[3], i.e. $\int_{\Gamma(t)} \mathrm{d}A \approx \int_{\mathcal{B}} \gamma_l(d) \mathrm{d}V$. However, the actual position and geometry of the crack have to be determined in advance by appropriate constraints of the type

$$D_{\Gamma(t)} = \{ d \,|\, d(\mathbf{x}, t) = 1 \text{ at } \mathbf{x} \in \Gamma(t) \}. \tag{2.25}$$

[3] The crack surface density $\gamma_l(d)$ is typically overestimated in our simulations. This partly results from the quadratic terms in its definition. Borden et al. [110] have shown that faster convergence of the approximation of the crack surface to its actual value can be achieved by an alternative formulation that takes into account higher gradients of the phase field. However, this poses certain requirements on the continuity of the approximative solution that cannot be implemented directly with a standard finite elements method. Nevertheless, in chapter 4 it is shown that the chosen formulation and implementation generate sufficiently accurate predictions of crack growth.

In order to obtain the full evolution of the growth of the crack, the minimization principle in equation (2.23), (2.24) and (2.25) is replaced by one involving the total energy. Borden et al. [117] present this in the form of a total Lagrangian energy functional

$$L(\boldsymbol{\varepsilon}, \dot{\mathbf{u}}, c, d) = \int_{\mathcal{B}} \mathrm{d}V \left(\frac{1}{2}\rho\dot{\mathbf{u}}^2 - \psi_{\mathrm{elastic}}(\boldsymbol{\varepsilon}, c, d) \right) - W_{\mathrm{crack}}(d), \qquad (2.26)$$

where ρ is the density of the material. The stored elastic energy density $\psi(\boldsymbol{\varepsilon}, c, d)$ is based on the constitutive law for an elastic linear isotropic material response summarized by equations (2.10 – 2.17). However, assuming that only tensile stress leads to crack growth, we decompose the energy density into compressive and tensile parts via

$$\psi^{\pm}(\boldsymbol{\varepsilon}, c) = \frac{1}{2}L\langle \mathrm{tr}[\boldsymbol{\varepsilon} - \boldsymbol{\varepsilon}_{\mathrm{Li}}]\rangle_{\pm}^{2} + G'\mathrm{tr}[\langle \boldsymbol{\varepsilon} - \boldsymbol{\varepsilon}_{\mathrm{Li}}\rangle_{\pm}^{2}]. \qquad (2.27)$$

Operating on scalars, the angle brackets are defined by $\langle x\rangle_{\pm} = \frac{1}{2}(|x| \pm x)$. Applied to tensor quantities, they project onto the eigenspace with positive or negative eigenvalues by $\langle \mathbf{A}\rangle_{\pm} = \sum_i \langle \alpha_i\rangle_{\pm}\, \boldsymbol{\nu}_i \otimes \boldsymbol{\nu}_i$, where α_i and $\boldsymbol{\nu}_i$ are the eigenvalues and corresponding eigenvectors of $\mathbf{A}$ respectively. The dyadic product $\mathbf{A} = \mathbf{v} \otimes \mathbf{w}$ is defined so that $A_{ij} = v_i w_j$, L is the Lamé constant and G' the shear modulus. The release of elastic energy at cracked locations is then modeled by multiplying the tensile part $\psi^{(+)}$ with a function, which becomes zero if the order parameter d approaches unity, so that

$$\psi_{\mathrm{elastic}}(\boldsymbol{\varepsilon}, c, d) = [(1-d)^2 + k]\, \psi^{+}(\boldsymbol{\varepsilon}, c) + \psi^{-}(\boldsymbol{\varepsilon}, c). \qquad (2.28)$$

Here, k is a numerical parameter chosen to stabilize the solution. The presented decomposition of ψ_{elastic} is exact in the sense that for $d = 0$ and $k = 0$, the ordinary isotropic linear elastic energy density is obtained again[4]. A further advantage of this formulation is that it naturally describes the behavior of crack faces under compressive force in a correct way. A numerically challenging tracking of possibly interpenetrating surface elements and the imposition of certain conditions to avoid non-physical behavior of the computational results, as often necessary with discrete descriptions of cracks, is therefore not required in our implemen-

[4] See Appendix A.2.

tation. Alternative approaches to distinguish between contributions to the elastic energy that come from different types of stress can be found in [112, 142, 143]. The Euler-Lagrange equations of the Lagrangian (2.26) are then used to find the evolution equation for the phase field

$$d - l^2 \nabla^2 d = 2(1-d)\frac{\mathcal{H}}{G_c/l}, \tag{2.29}$$

where the local history field of maximum positive elastic stored energy

$$\mathcal{H}(\mathbf{x}, t) := \max_{s \in [0,t]} \psi^+(\boldsymbol{\varepsilon}(\mathbf{x}, s), c(\mathbf{x}, s)), \tag{2.30}$$

replaces $\psi^+(\boldsymbol{\varepsilon})$ to ensure that the crack growth is irreversible so that no crack healing can occur. In equation (2.30), s describes all points in time from 0 to the current one t. The elastic energy that drives the crack results from conservation of linear momentum as stated in equation (2.7), consistent with the Euler-Lagrange equations.

The coupled pair of balance equations (2.7) and (2.29), together with the constitutive relations for the Cauchy stress resulting from the corresponding partial derivative of $\psi_{\text{elastic}}(\boldsymbol{\varepsilon}, d)$, as well as the definition of the history field (2.30) and the kinematic relation for the strain (2.8), represent a well-defined mathematical problem. The necessary boundary conditions to this problem are given by $\mathbf{u} = \bar{\mathbf{u}}$ on ∂B_u and $\mathbf{t} = \boldsymbol{\sigma}\mathbf{n} = \bar{\mathbf{t}}$ on ∂B_t, as stated above, as well as $\nabla d \cdot \mathbf{n} = 0$ on ∂B.

2.3 Fracture in a Storage Particle

We assume, that the driving force for fracture is solely generated by the tensile elastic energy present in the material. In this approximation, chemical effects of the Li on the initiation and propagation of a crack are neglected and only its effect of swelling the material is accounted for. Alternative formulations, considering coupled mechano-chemical driving forces for fracture, can be found in [144, 145]. However, the simulation of the growth of a crack is not pursued in those works.

By virtue of the formulation via a single free energy density, it is a straightforward matter to couple the stress diffusion model to the phase field describing crack growth. The central quantity that links the different fields is the elastic energy. Under the assumption, that only mechanical energy from elastic strain induces fracture, the spectral decomposition in equations (2.27) and (2.28) of section 2.2, is transfered to the full free energy density of the stress diffusion model (2.14), together with the fracture energy, yielding a replacement for equation (2.14) as

$$\psi(\boldsymbol{\varepsilon}, c, d) = [(1-d)^2 + k]\, \psi^+(\boldsymbol{\varepsilon}, c) + \psi^-(\boldsymbol{\varepsilon}, c) + G_c \gamma_l(d) - R\theta \int_0^c \mathrm{d}c' \ln \frac{1 - c'/c_{\max}}{c'/c_{\max}} + \mu_0 c. \tag{2.31}$$

The functional dependence of the free energy density on the strain is not affected by the spectral decomposition in the sense, that $\psi(\boldsymbol{\varepsilon}, c, d)$ is still only dependent on the elastic strain $\boldsymbol{\varepsilon}_{\text{elastic}} = \boldsymbol{\varepsilon} - \boldsymbol{\varepsilon}_{\text{Li}}$. As a consequence, the form of the partial derivative with respect to the concentration remains unaltered when expressed through the hydrostatic stress, as defined in equation (2.17), i.e.

$$\frac{\partial \psi_{\text{elastic}}}{\partial c} = \frac{\partial \psi_{\text{elastic}}}{\partial(\boldsymbol{\varepsilon} - \boldsymbol{\varepsilon}_{\text{Li}})} : \frac{\partial(\boldsymbol{\varepsilon} - \boldsymbol{\varepsilon}_{\text{Li}})}{\partial c} = -\frac{\Omega}{3}\boldsymbol{\sigma} : \mathbf{1} = -\frac{\Omega}{3}\mathrm{tr}[\boldsymbol{\sigma}] = -\Omega\, \sigma_h. \tag{2.32}$$

The constitutive equation for the flux, as expressed in equation (2.20), therefore remains valid in the coupled model. However, since the spectral decomposition influences the Cauchy stress $\boldsymbol{\sigma}$, the hydrostatic stress entering the flux is modified. The Cauchy stress $\boldsymbol{\sigma}$ is again obtained through a partial derivative of $\psi(\boldsymbol{\varepsilon}, c, d)$ as

$$\boldsymbol{\sigma} = \partial_{\boldsymbol{\varepsilon}} \psi = [(1-d)^2 + k]\, \partial_{\boldsymbol{\varepsilon}} \psi^+ + \partial_{\boldsymbol{\varepsilon}} \psi^-) = [(1-d)^2 + k]\, \boldsymbol{\sigma}^+ + \boldsymbol{\sigma}^-, \tag{2.33}$$

with the compressive and tensile stress given by

$$\boldsymbol{\sigma}^{\pm} = L \langle \mathrm{tr}[\boldsymbol{\varepsilon} - \boldsymbol{\varepsilon}_{\text{Li}}] \rangle_{\pm} \mathbf{1} + G' \langle \boldsymbol{\varepsilon} - \boldsymbol{\varepsilon}_{\text{Li}} \rangle_{\pm}. \tag{2.34}$$

In this first approximation, the presence of a crack affects the diffusion of Li only through the modification in the stress field. A direct impact of the crack on the Li flux by hindering or blocking flow perpendicular to the crack faces, enhancing or mitigating diffusion along the faces, or introducing additional flux through the faces, is not examined in this work and remains as a future task.

For the sake of a better overview, we summarize the relevant equations in FIG. 2.2. The initial and boundary conditions of the coupled model are directly taken over from section 2.1 for the displacement and concentration field $\mathbf{u}$ and c and from section 2.2 for the phase field d.

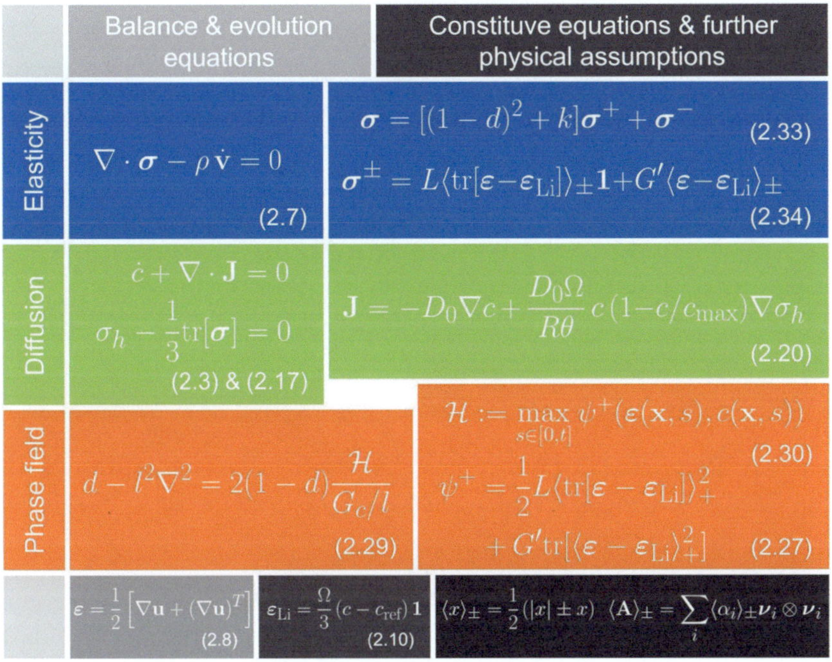

Figure 2.2.: Summary of the derived partial differential and algebraic equations. The model describes Li diffusion, mechanical stress and crack growth in a simultaneous fashion. The different physical blocks are mutually coupled and demand for an elaborated solving strategy (see FIG. 3.2).

3 Numerical Implementation

3.1 From Differential To Matrix Equations

Due to the complexity of the model derived in the previous chapter, the system of partial differential equations (PDE) presented in FIG. 3.2, can only be solved using an appropriate numerical method such as the finite element method (FEM). To this purpose, the boundary value problem (BVP) has to be cast in a suitable form, so that it can be implemented in a computer algorithm. We describe the steps required to obtain this form in the following sections.

3.1.1 Time Discretization

Various approaches exist for numerical solution of a boundary value problem, which is both depending on time and space. For example, one may first discretize the equations in space and then solve the remaining ordinary differential equation in time. This approach is generally referred to as the so called method of lines [146] and works particularly well for parabolic partial differential equations. However, for the elliptic equation describing the balance of linear momentum, it is more common to resolve first the differential operator in time by a suitable time integration method. Thereafter the resulting equation is solved for its spatial variables. In our work, the implicit Backward-Euler time integration method is applied. Alternatives, such as, for example, the explicit Forward-Euler or Crank-Nicholson method were tested, but showed significant disadvantages in terms of stability.

We divide the observation time T into $\mathcal{S}$ time steps and impose as requirements on the time increments that $\Delta t_n \in \mathbb{R}^+$, $t_{n+1} = t_n + \Delta t_{n+1}$ and $T = t_{\mathcal{S}-1}$, where $n \in \{0; 1; ...; \mathcal{S} - 1\}$. Applying the Backward-Euler method for the time step $n+1$, the balance of Li content (2.3) then reads

$$\frac{c_{n+1}(\mathbf{x})}{\Delta t_{n+1}} - \frac{c_n(\mathbf{x})}{\Delta t_{n+1}} + \nabla \cdot \mathbf{J}_{n+1}(\mathbf{x}) = 0. \tag{3.1}$$

With the concentration at the previous step $c_n(\mathbf{x})$ being given, one thus has to solve for the current concentration $c_{n+1}(\mathbf{x})$. The flux is hereby computed from the current solution. This is different in explicit methods, where the flux is computed from the previous solution. The time integration eliminates the derivative with respect to time. Hence, the remaining differential operators act solely on the spatial variable. For the sake of brevity, we drop the subscript for the current time step and also the dependence on the spatial location to give

$$\frac{c}{\Delta t_{n+1}} - \frac{c_n}{\Delta t_{n+1}} + \nabla \cdot \mathbf{J} = 0. \tag{3.2}$$

The reason for keeping the subscript at the time increment Δt_{n+1} is not to be confused with time increments of previous steps, which may be different due to the time-adaptive method implemented in our algorithm.

The necessity of working with a time-adaptive method can be understood by examination of the characteristic time scales on which the different physical phenomena of the model take place. On the one hand, the diffusive motion of the Li-ions is a comparatively slow process, determined by the mobility of the ions and the net relative distance they have to travel. Performing a simple dimensional analysis, we obtain a characteristic time of $\tau_{\text{diffusion}} = r_0^2/D_0$, where r_0 is the typical length scale of the storage particle in question, e.g. the radius of a spherical particle. We set it to $r_0 = 1\,\mu\text{m}$ throughout this work. With diffusion coefficients around $D_0 \approx 10^{-14}\,\text{m}^2/\text{s}$, this gives $\tau_{\text{diffusion}} \approx 10^2\,\text{s}$. On the other hand, where inertia effects dominate the elastic behavior of the material, e.g. during unstable crack growth, action takes place on a much shorter time scale. A rough estimate gives the time that a sound wave needs to travel from one side of the particle to the other. If $v_{\text{wave}} = \sqrt{E/\rho}$ is approximately the speed of sound in matter, the characteristic elastic time scale is $\tau_{\text{elastic}} = r_0/v_{\text{wave}} = r_0\sqrt{\rho/E}$. With typical values of $\rho \approx 10^3\,\text{kg}/\text{m}^3$ for the density and $E \approx 100\,\text{GPa}$ for Young's modulus, this yields $\tau_{\text{elastic}} \approx 10^{-10}\,s$. With respect to the ratio of these characteristic time scales $\tau_{\text{diffusion}}/\tau_{\text{elastic}} \approx 10^{12}$, it becomes obvious, that in order to monitor each physical process adequately, an elaborate time adaptation method is necessary. This will be presented in section 3.3.4.

The variable time increment must be taken into account in the application of the time integration method to the equation describing the balance of linear momentum (2.7). Proceeding in a similar way as above, we obtain

$$\frac{\rho\,\mathbf{u}}{\Delta t_{n+1}^2} - \frac{\rho\,\mathbf{u}_n}{\Delta t_{n+1}^2} - \frac{\rho\,(\mathbf{u}_n - \mathbf{u}_{n-1})}{\Delta t_{n+1}\Delta t_n} = \nabla\cdot\boldsymbol{\sigma}, \tag{3.3}$$

where we have used $\dot{\mathbf{v}}$ and $\dot{\mathbf{u}} = \mathbf{v}$. With displacement fields from the last two time steps $\mathbf{u}_n$ and $\mathbf{u}_{n-1}$ being known, we solve for $\mathbf{u}$ and compute $\boldsymbol{\sigma}$ from the current solution.

3.1.2 Spatial Discretization

After execution of the time integration method, the resulting partial differential equations only involve differential operators acting on spatial variables. In order to obtain solutions for the relevant fields at each time step, we employ the so called Galerkin's method, a scheme that is commonly used in the finite element method [147]. It transfers the continuous differential operator problem to a discrete matrix problem. We come back to the the different steps involved in Galerkin's method and explain them in detail after first extending the system of partial differential equations by an additional equation in the sense of a mixed formulation.

Mixed Formulation

A set of differential equations that depends on the variables $a_1, ..., a_N$, which is expressed by

$$\begin{aligned} \mathcal{A}(a_1, ..., a_N) &= \mathbf{0} \qquad && \text{in the domain } \mathcal{B} \text{ and} \\ \mathcal{B}(a_1, ..., a_N) &= \mathbf{0} && \text{on its boundary } \partial\mathcal{B}, \end{aligned}$$

is formulated in a so called irreducible way if none of the variables $a_1, ..., a_N$ can be eliminated while leaving the problem in a well-defined way. If this is not the case, the formulation is called mixed [147]. Mixed formulations can effect the requirement on the continuity of the shape functions, approximating the solution of the problem, and may stabilize the solution of numerically delicate problems. They have been employed successfully in, for example, problems with almost incompressible elastic bodies or fluid dynamics, such as the Stokes equation [148, 149]. Here, we use a mixed formulation that defines the hydrostatic stress as an additional independent variable, as given in equation (2.17), through

$$\sigma_h - \frac{1}{3}\mathrm{tr}[\boldsymbol{\sigma}] = 0. \tag{3.4}$$

This approach has proven to be numerically advantageous in our implementation. We attribute this to the smoothness of the numerical solution in a standard finite element method. As explained on page 37, the numerical approximation to the current solution is continuous, but not continuously differentiable. Spatial derivatives of the approximate solution, in particular those of higher order, are thus allowed to jump. In our model, this concerns the hydrostatic stress, which enters the equation for balance of Li in the form of its spatial gradient. In the mixed formulation, the second order derivatives of the displacement field are transferred to first order derivatives of a field, that on the other hand includes first order derivatives of the displacement field. In the numerical implementation, this leads to a smoother solution since jumps of the first order derivatives are replaced by their averaged values. This results in a better convergence of our numerical scheme.

Weak Form

The translation of the PDE to matrix equations, as pursued in Galerkin's method, commences with the expression of the corresponding boundary value problem in its weak form. We therefore multiply the balance equations with so called variational test functions δc, $\delta \mathbf{u}$ and δd and integrate over the full body. For the equation describing balance of Li content (3.2), we obtain

$$\begin{aligned} G_c(c, d, \sigma_h, \delta c) = & \int_{\mathcal{B}} \mathrm{d}V \frac{c}{\Delta t_{n+1}}\,\delta c - \int_{\mathcal{B}} \mathrm{d}V \frac{c_n}{\Delta t_{n+1}}\,\delta c \\ & + \int_{\mathcal{B}} \mathrm{d}V\, \nabla \cdot \mathbf{J}\, \delta c = 0, \end{aligned} \tag{3.5}$$

where the dependence of $G_c(\mathbf{u}, c, d, \delta c)$ on the displacement and phase field is hidden in the flux. The variational test functions are members of a certain function space consistent with the given boundary conditions and solutions of (3.5) are so called weak solutions of (3.2) with respect to the chosen set of test functions. For arbitry δc, it can be shown that equations (3.2) and (3.5) are equivalent. Using integration by parts we rearrange the weak form to

$$\begin{aligned} G_c(c, d, \sigma_h, \delta c) = & \int_{\mathcal{B}} \mathrm{d}V \frac{c}{\Delta t_{n+1}} \delta c - \int_{\mathcal{B}} \mathrm{d}V \frac{c_n}{\Delta t_{n+1}} \delta c \\ & - \int_{\mathcal{B}} \mathrm{d}V \, \mathbf{J} \cdot (\nabla \delta c) + \int_{\partial\mathcal{B}} \mathrm{d}S \, \bar{\mathbf{J}} \cdot \mathbf{n} \, \delta c = 0, \end{aligned} \tag{3.6}$$

Following the same procedure for the equation describing the balance of linear momentum yields

$$\begin{aligned} G_{\mathbf{u}}(\mathbf{u}, c, d, \delta \mathbf{u}) = & \int_{\mathcal{B}} \mathrm{d}V \frac{\rho \, \mathbf{u}}{\Delta t_{n+1}^2} \delta \mathbf{u} - \int_{\mathcal{B}} \mathrm{d}V \frac{\rho \, \mathbf{u}_n}{\Delta t_{n+1}^2} \delta \mathbf{u} + \int_{\mathcal{B}} \mathrm{d}V \boldsymbol{\sigma} : \delta \boldsymbol{\varepsilon} \\ & - \int_{\mathcal{B}} \mathrm{d}V \frac{\rho \, (\mathbf{u}_n - \mathbf{u}_{n-1})}{\Delta t_{n+1} \Delta t_n} \delta \mathbf{u} - \int_{\partial\mathcal{B}} \mathrm{d}S \, \bar{\mathbf{t}} \cdot \delta \mathbf{u} = 0, \end{aligned} \tag{3.7}$$

where the symmetry relation $\boldsymbol{\sigma} : \nabla(\delta \mathbf{u}) = \boldsymbol{\sigma} : \frac{1}{2}[\nabla(\delta \mathbf{u}) + (\nabla(\delta \mathbf{u}))^T] = \boldsymbol{\sigma} : \delta \boldsymbol{\varepsilon}$ was exploited. The dependence of $G_u(\mathbf{u}, c, d, \delta \mathbf{u})$ on the concentration and the phase field results from the definition of the stress tensor. The weak form of the evolution equation for the phase field reads

$$\begin{aligned} G_d(d, \mathbf{u}, c, \delta d) = & \int_{\mathcal{B}} \mathrm{d}V \left(\frac{G_c}{l} + 2\mathcal{H} \right) d \, \delta d \\ & + \int_{\mathcal{B}} \mathrm{d}V \, G_c \, l \, \nabla d \cdot \nabla(\delta d) - 2 \int_{\mathcal{B}} \mathrm{d}V \mathcal{H} \, \delta d = 0, \end{aligned} \tag{3.8}$$

which depends on the displacement and concentration field through the history field $\mathcal{H}$. For the additional equation resulting from the mixed formulation, we obtain

$$G_{\sigma_h}(\mathbf{u}, c, d, \sigma_h, \delta \sigma_h) = \int_{\mathcal{B}} \mathrm{d}V \sigma_h \delta \sigma_h - \int_{\mathcal{B}} \mathrm{d}V \frac{1}{3} \mathrm{tr} \boldsymbol{\sigma} \, \delta \sigma_h = 0. \tag{3.9}$$

As before, the dependence on the displacement field, concentration and phase field is contained in the definition of the stress.

Staggered Solution Scheme

In order to obtain an effective algorithm in terms of computation time and memory consumption, we aim for solving the system of partial differential equations in a staggered scheme. Among other advantages, such as a simplified implementation, this allows to exploit specific properties of the partial differential equations, such as symmetry and positive definiteness. These properties are reflected in the corresponding matrix equations, presented later in this chapter. The staggered approach thus permits the choice of numerical solvers that are most suitable for each equation[1]. In contrast to a monolithic approach, where the full system is solved as a whole, a staggered approach solves only parts of the system while holding others fixed. As an example, the equation describing conservation of linear momentum is solved for a fixed concentration field. A drawback of this approach is that increments of solutions between different time steps have to remain small. This is granted by the time-adaptive method presented in section 3.3.4. In the derived weak forms, most of the variables thus only act as parameters. The distinction between the variables for which the respective weak form is solved and those that are treated as parameters is formally represented by a vertical bar in the following notation

$$G_c(c, \sigma_h, \delta c) \rightarrow G_c(c, \delta c \,|\, \sigma_h), \tag{3.10}$$

$$G_{\mathbf{u}}(\mathbf{u}, c, d, \delta \mathbf{u}) \rightarrow G_{\mathbf{u}}(\mathbf{u}, \delta \mathbf{u} \,|\, c, d), \tag{3.11}$$

$$G_d(d, \mathbf{u}, c, \delta d) \rightarrow G_d(d, \delta d \,|\, \mathbf{u}, c), \tag{3.12}$$

$$G_{\sigma_h}(\mathbf{u}, c, d, \sigma_h, \delta \sigma_h) \rightarrow G_{\sigma_h}(\sigma_h, \delta \sigma_h \,|\, \mathbf{u}, c, d). \tag{3.13}$$

Remarks on Linearization

Seeking for solutions of equations (3.6 – 3.9) requires an appropriate solution scheme. The standard algorithm in the field of the finite element method is given by the Newton-Raphson method. Its basic idea is equivalent to Newton's method for an ordinary scalar equation of the type

$$f(x) = 0, \tag{3.14}$$

[1] Implementation of the equations in a single monolithic matrix would make it mandatory to use a direct solver. The scaling behavior of these solvers with respect to the size of the system is generally significantly worse than for iterative solvers. Hence, detailed results as shown, for example, in the study of Li insertion in chapter 6, would not have been achievable in this approach.

where $f(x)$ is steady and differentiable in a region X around the solution x^*. Starting with an initial guess $x_m \in X$ that is close enough to x^*, it is straight-forward to show that the solution is approached iteratively via

$$x_{m+1} = x_m - \frac{f(x_m)}{f'(x_m)}. \tag{3.15}$$

In the language of the finite element method $f(x_m)$ and $f'(x_m)$ are labeled residuum $\mathbf{R}$ and tangent operator $\mathbf{K}$ respectively. In order to obtain the equivalent of $\mathbf{K}$ for the above derived weak forms, one has to compute the partial derivatives with respect to the relevant unknown fields, i.e. in our case $\mathbf{u}$, c, d and σ_h. Technically, this is calculated by the directional derivative (cf. [150]), which for the functional G_u and with respect to the variable $\mathbf{u}$ has the following form

$$D_{\mathbf{u}}[G_{\mathbf{u}}(\mathbf{u}, c, d, \delta\mathbf{u})]_{|\mathbf{u}=\mathbf{u}_0} = \frac{\mathrm{d}}{\mathrm{d}\epsilon} G_{\mathbf{u}}(\mathbf{u}_0 + \epsilon\Delta\mathbf{u}, c, d, \delta\mathbf{u})_{|\epsilon=0}. \tag{3.16}$$

In particular, the dependence of $G_{\mathbf{u}}$ on one of the independent variables includes also their spatial derivatives. We note that within the theory of infinitesimal strain, the linear change of an independent variable, such as for example $\Delta\mathbf{u}$, commutes with the spatial derivative of the same variable $\nabla\mathbf{u}$, i.e. $\Delta(\nabla\mathbf{u}) = \nabla(\Delta\mathbf{u})$. This relation enables us to perform the linearization with a formal chain rule like procedure, i.e.

$$D_{\mathbf{u}}[G] = \Delta_{\mathbf{u}}[G] + \Delta_{\nabla\mathbf{u}}[G] + \ldots = \frac{\partial G}{\partial \mathbf{u}} \cdot (\Delta\mathbf{u}) + \frac{\partial G}{\partial(\nabla\mathbf{u})} : \nabla(\Delta\mathbf{u}) + \ldots$$

In order to keep the notation tight, we further use the relation

$$\Delta_{\nabla\mathbf{u}}[G] = \frac{\partial G}{\partial \boldsymbol{\varepsilon}} \Delta\boldsymbol{\varepsilon}, \tag{3.17}$$

where the linear increment of $\boldsymbol{\varepsilon}$ is given by

$$\Delta\boldsymbol{\varepsilon} = \Delta_{\mathbf{u}}[\boldsymbol{\varepsilon}] = \Delta_{\mathbf{u}}[\mathrm{sym}[\nabla\mathbf{u}]] = \mathrm{sym}[\nabla(\Delta\mathbf{u})]. \tag{3.18}$$

Linearization

Due to the staggered solution scheme and the resulting distinction between independent variables and constant parameters, each of the weak forms given in equations (3.6 – 3.9) is only linearized with respect to the respective independent variable. This simplifies the linearization procedure significantly. The linearized weak forms therefore read

$$\mathrm{Lin}[G_\square] = G_\square + D_\square[G_\square], \tag{3.19}$$

with the square symbol $\square$ representing the four fields c, $\mathbf{u}$, d and σ_h and the linear increments given by

$$D_c[G_c] = \int_{\mathcal{B}} \mathrm{d}V \frac{1}{\Delta t_{n+1}} \delta c \, \Delta c - \int_{\mathcal{B}} \mathrm{d}V (\nabla \delta c) \cdot (\mathbf{J}_c \Delta c + \mathbf{J}_{\nabla c} \cdot (\nabla \Delta c)), \tag{3.20}$$

$$D_{\mathbf{u}}[G_{\mathbf{u}}] = \int_{\mathcal{B}} \mathrm{d}V \frac{\rho}{\Delta t_{n+1}^2} \delta \mathbf{u} \cdot \Delta \mathbf{u} + \int_{\mathcal{B}} \mathrm{d}V \delta \boldsymbol{\varepsilon} \colon \mathbb{C} \colon \Delta \boldsymbol{\varepsilon}, \tag{3.21}$$

$$D_d[G_d] = \int_{\mathcal{B}} \mathrm{d}V \left(\frac{G_c}{l} + 2\mathcal{H} \right) \delta d \, \Delta d + \int_{\mathcal{B}} \mathrm{d}V G_c l \, (\nabla \delta d) \cdot (\nabla \Delta d), \tag{3.22}$$

$$D_{\sigma_h}[G_{\sigma_h}] = \int_{\mathcal{B}} \mathrm{d}V \, \delta \sigma_h \, \Delta \sigma_h, \tag{3.23}$$

in terms of the linear increments of the concentration Δc, displacements $\Delta \mathbf{u}$, phase field Δd and hydrostatic stress $\Delta \sigma_h$ respectively. The partial derivatives of the flux with respect to c and ∇c read

$$\mathbf{J}_c := \partial \mathbf{J} / \partial c = \frac{D_0 \Omega}{R\theta} (1 - 2c/c_{\max}) \nabla \sigma_h, \tag{3.24}$$

$$\mathbf{J}_{\nabla c} := \partial \mathbf{J} / \partial (\nabla c) = -D_0 \mathbf{1}, \tag{3.25}$$

and the forth order stiffness tensor is defined by $\mathbb{C} := \partial \boldsymbol{\sigma} / \partial \boldsymbol{\varepsilon}$. We follow a similar approach as in [151] and express it in terms of eigenvalues λ_i and eigenvectors $\boldsymbol{\nu}_i$ of the strain tensor by

$$\mathbb{C} = \sum_{a,b=1}^{3} \frac{\partial s_b}{\partial \lambda_a} \boldsymbol{\nu}_b \otimes \boldsymbol{\nu}_b \otimes \boldsymbol{\nu}_a \otimes \boldsymbol{\nu}_a + \frac{1}{2} \sum_{\substack{a,b=1 \\ a \neq b}} \frac{s_b - s_a}{\lambda_b - \lambda_a} \left(\boldsymbol{\nu}_a \otimes \boldsymbol{\nu}_b \otimes \boldsymbol{\nu}_a \otimes \boldsymbol{\nu}_b + \boldsymbol{\nu}_a \otimes \boldsymbol{\nu}_b \otimes \boldsymbol{\nu}_b \otimes \boldsymbol{\nu}_a \right), \tag{3.26}$$

with s_i denoting an eigenvalue of the Cauchy stress tensor. Due to the isotropy of the free energy density potential, the eigenvectors of the stress and strain tensor are co-linear and the eigenvalues can be expressed as $s_i = \partial\psi/\partial\lambda_i$. Inserting (2.28), we express the respective terms by

$$\frac{\partial s_b}{\partial \lambda_a} = [(1-d)^2 + k] \left[L \frac{\langle \sum_{j=1}^{3} \lambda_j \rangle_+}{\sum_{j=1}^{3} \lambda_j} + 2G'\,\delta_{ab} \frac{\langle \lambda_b \rangle_+}{\lambda_b} \right] + \left[L \frac{\langle \sum_{j=1}^{3} \lambda_j \rangle_-}{\sum_{j=1}^{3} \lambda_j} + 2G'\,\delta_{ab} \frac{\langle \lambda_b \rangle_-}{\lambda_b} \right], \tag{3.27}$$

$$\frac{s_b - s_a}{\lambda_b - \lambda_a} = [(1-d)^2 + k]\, 2G' \frac{\langle \lambda_b \rangle_+ - \langle \lambda_a \rangle_+}{\lambda_b - \lambda_a} + 2G' \frac{\langle \lambda_b \rangle_- \langle \lambda_a \rangle_-}{\lambda_b - \lambda_a}. \tag{3.28}$$

In cases where eigenvalues are zero or equal to each other, the above expressions are understood in an asymptotic sense. Namely, it can be shown that as a denominator approaches zero, the corresponding enumerator tends to zero at the same rate. Each of the terms in equations (3.26 – 3.28) is therefore well-defined and no singularities occur.

Approximation through Shape Functions

In the finite element method, the domain representing the geometry of the particle is described by a mesh of $\mathcal{M}$ elements. Depending on the type of element, they are comprised of a number of nodes that either lie on corners, edges, faces or inside the element. The actual fields, e.g. the concentration field c, are then replaced by an approximation c^h, which is assumed to be a linear combination of so called basis or shape functions

$$c \approx c^h = \sum_{j}^{\mathcal{N}} N_j(\mathbf{x}) D_j^c = \mathbf{N}\mathbf{D}^c, \tag{3.29}$$

where the index j runs over the total number of nodes $\mathcal{N}$. The short notation on the right hand side is understood in a vector-like sense as $\mathbf{N} = [N_1\, N_2 \ldots N_{\mathcal{N}}]$ and $\mathbf{D}^c = [D_1^c\, D_2^c \ldots D_{\mathcal{N}}^c]^T$. For numerical reasons, the shape functions are defined such that each of them becomes unity at the location of one node and zero at all other nodes[2]. If $\mathbf{x}_i$ is the location of the i-th node, this means that $N_j(\mathbf{x}_i) = \delta_{ij}$. Hence, D_i represents the

[2] The advantage of this definition lies in the sparsity of the resulting matrix equation.

value of the approximated solution at the point $\mathbf{x}_i$. Further, nodes adjacent to two or more elements are shared between the respective elements which implies C^0-continuity of the approximate solution. We proceed in the same way for the displacements, phase field and hydrostatic stress[3]

$$\mathbf{u} \approx \mathbf{u}^h = \sum_j^{\mathcal{N}} N_j(\mathbf{x})\mathbf{D}_j^{\mathbf{u}} = \mathbf{N}\mathbf{D}^{\mathbf{u}}, \tag{3.30}$$

$$d \approx d^h = \sum_j^{\mathcal{N}} N_j(\mathbf{x})D_j^d = \mathbf{N}\mathbf{D}^d, \tag{3.31}$$

$$\sigma_h \approx \sigma_h^h = \sum_j^{\mathcal{N}} N_j(\mathbf{x})D_j^{\sigma_h} = \mathbf{N}\mathbf{D}^{\sigma_h}, \tag{3.32}$$

with the short notations[4] $\mathbf{D}^{\mathbf{u}} = [\mathbf{D}_1^{\mathbf{u}}\,\mathbf{D}_2^{\mathbf{u}}\ldots\mathbf{D}_{\mathcal{N}}^{\mathbf{u}}]^T, \mathbf{D}^d = [D_1^d\,D_2^d\ldots D_{\mathcal{N}}^d]^T$ and $\mathbf{D}^{\sigma_h} = [D_1^{\sigma_h}\,D_2^{\sigma_h}\ldots D_{\mathcal{N}}^{\sigma_h}]^T$. As apparent from the definitions given in equations (3.29 – 3.32), the spatial dependence of the approximate solutions is solely covered by the shape functions. This allows to express the differential operators appearing in the weak forms by simple matrix-vector products. For example, replacing c by its approximation c^h, the gradient of the concentration field becomes

$$\nabla c \approx \nabla c^h = \nabla(\mathbf{N}\mathbf{D}^c) = (\nabla^T\mathbf{N})\mathbf{D}^c = \mathbf{B}\mathbf{D}^c \tag{3.33}$$

with the $3 \times \mathcal{N}$ matrix $B_{ij} = \partial N_i/\partial x_j$. In a similar way, we substitute the strain tensor via

$$\boldsymbol{\varepsilon} = \mathrm{sym}[\nabla\mathbf{u}] \approx \mathrm{sym}[\nabla\mathbf{u}^h] = \sum_j^{\mathcal{N}} \mathrm{sym}[N_j(\mathbf{x})\mathbf{D}_j^{\mathbf{u}}] = \sum_j^{\mathcal{N}} \mathbf{S}^j\mathbf{D}_j^{\mathbf{u}} = \mathbf{S}\mathbf{D}^{\mathbf{u}}, \tag{3.34}$$

[3] It is implicit in these definitions, that only one type of shape functions (Quadratic Lagrange functions) is used for all the relevant physical fields. Working with shape functions of different orders has been tested in a small number of simulations, but showed no significant improvement in terms of computation time.

[4] For the sake of a breve notation, we define $\mathbf{D}^{\mathbf{u}}$ formally by a ordered representation of vectors. In general, vector-valued problems, like the momentum equation, require a specific enumeration of the nodal values for an actual implementation in a computer code. A typical scheme for the displacement field is to number coefficient D_i describing x-displacements with $i = 3j$, y-displacements with $i = 3j + 1$ and z-displacements with $i = 3j + 2$, where j is the corresponding node number.

where the third order operator $\mathbf{S}^j$ is given by $S^j_{klm} = \delta_{km}\partial_l N_j + \delta_{lm}\delta_k N_j$ with $k, m, l \in \{0, 1, 2\}$ and $j \in \{0, 1, \dots, \mathcal{N}\}$. A last definition shall be made to define the dyadic product of the vectors of shape functions, often referred to as mass matrix at unit density

$$\mathbf{M} = \mathbf{N}^T \mathbf{N}. \tag{3.35}$$

The above definitions apply in the same way to the test functions and the linear increments. While the shape functions are unaffected, the nodal values are replaced by the corresponding virtual displacements and linear increments respectively. For the concentration this yields, for example,

$$\Delta c \approx \Delta c^h = \sum_j^{\mathcal{N}} N_j(\mathbf{x}) \Delta D^c_j = \mathbf{N}\, \Delta \mathbf{D}^c, \tag{3.36}$$

$$\delta c \approx \delta c^h = \sum_j^{\mathcal{N}} N_j(\mathbf{x}) \delta D^c_j = \mathbf{N}\, \delta \mathbf{D}^c. \tag{3.37}$$

We mentioned earlier that the test functions lie in a certain function space. By equation (3.37) we assume, that the shape functions lie in the same function space. The choice of the same set of shape functions for both δc and c or Δc is called symmetric Galerkin's method. We insert equation (3.29), (3.36) and (3.37) into the linearized weak form of equation (3.19), with the linear increment given by (3.20), yielding

$$\begin{aligned}
\mathrm{Lin}[G^h_c] = & \int_{\mathcal{B}} \mathrm{d}V \frac{1}{\Delta t_{n+1}} \delta \mathbf{D}^{c^T} \mathbf{M}\, \mathbf{D}^c - \int_{\mathcal{B}} \mathrm{d}V \frac{1}{\Delta t_{n+1}} \delta \mathbf{D}^{c^T} \mathbf{M}\, \mathbf{D}^c_n \\
& - \int_{\mathcal{B}} \mathrm{d}V \delta \mathbf{D}^{c^T} \mathbf{B}^T \mathbf{J}^h - \int_{\partial \mathcal{B}_J} \mathrm{d}S\, \delta \mathbf{D}^{c^T} \mathbf{N}^T\, (\bar{\mathbf{J}}^h \cdot \mathbf{n}) \\
& + \int_{\mathcal{B}} \mathrm{d}V \frac{1}{\Delta t_{n+1}} \delta \mathbf{D}^{c^T} \mathbf{M}\, \Delta \mathbf{D}^c - \int_{\mathcal{B}} \mathrm{d}V \delta \mathbf{D}^{c^T} \mathbf{B}^T \mathbf{J}^h_c\, \mathbf{N}\, \Delta \mathbf{D}^c \\
& - \int_{\mathcal{B}} \mathrm{d}V \delta \mathbf{D}^{c^T} \mathbf{J}^h_{\nabla c} \mathbf{B}^T \mathbf{B}\, \Delta \mathbf{D}^c = 0.
\end{aligned} \tag{3.38}$$

Terms marked with the superscript 'h' must be understood as approximations to the actual parameters. For example, $\mathbf{J}^h$ is the Li flux computed via (2.20) using the approximate concentration field c^h.

As mentioned earlier, the initial weak form presented in equation (3.5), from which the linearized, approximated expression of equation (3.38) is derived, needs to be fulfilled for an arbitrary choice of test functions δc, living in a certain function space. This principle transfers to the linearized expression in the sense that it needs to be fulfilled for an arbitrary choice of $\delta \mathbf{D}^c$. By successively setting, for example, all but one component of $\delta \mathbf{D}^c$ to zero, expression (3.38) yields $\mathcal{N}$ independent equations. Casting these into the form of a matrix equation, we obtain

$$\mathbf{R}_c + \mathbf{K}_{cc} \Delta \mathbf{D}^c = \mathbf{0}, \tag{3.39}$$

with the residual vector and the tangent matrix

$$\begin{aligned} \mathbf{R}_c = & \int_{\mathcal{B}} \mathrm{d}V \frac{1}{\Delta t_{n+1}} \mathbf{M}\,\mathbf{D}^c - \int_{\mathcal{B}} \mathrm{d}V \frac{1}{\Delta t_{n+1}} \mathbf{M}\,\mathbf{D}^c_n \\ & - \int_{\mathcal{B}} \mathrm{d}V \mathbf{B}^T \mathbf{J}^h - \int_{\partial \mathcal{B}_J} \mathrm{d}S\,\mathbf{N}^T \left(\bar{\mathbf{J}}^h \cdot \mathbf{n} \right), \end{aligned} \tag{3.40}$$

$$\mathbf{K}_{cc} = \int_{\mathcal{B}} \mathrm{d}V \frac{1}{\Delta t_{n+1}} \mathbf{M} - \int_{\mathcal{B}} \mathrm{d}V \mathbf{B}^T \mathbf{J}^h_c\,\mathbf{N} - \int_{\mathcal{B}} \mathrm{d}V\,J^h_{\nabla c}\,\mathbf{B}^T \mathbf{B}. \tag{3.41}$$

The continuous boundary value problem is thus transformed to a matrix equation and the remaining difficulty of inverting the $\mathcal{N} \times \mathcal{N}$ tangent matrix $\mathbf{K}_{cc}$ can be forwarded to a computer algorithm. We perform the same steps to receive the matrix equations for the displacement, phase and hydrostatic stress field

$$\mathbf{R}_{\mathbf{u}} + \mathbf{K}_{\mathbf{uu}} \Delta \mathbf{D}^{\mathbf{u}} = \mathbf{0}, \tag{3.42}$$

$$\mathbf{R}_d + \mathbf{K}_{dd} \Delta \mathbf{D}^d = \mathbf{0}, \tag{3.43}$$

$$\mathbf{R}_{\sigma_h} + \mathbf{K}_{\sigma_h \sigma_h} \Delta \mathbf{D}^{\sigma_h} = \mathbf{0}, \tag{3.44}$$

where the residuals are expressed by

$$\begin{aligned} \mathbf{R}_{\mathbf{u}} = & \int_{\mathcal{B}} \mathrm{d}V \frac{\rho}{\Delta t^2_{n+1}} \mathbf{M} \left(\mathbf{D}^{\mathbf{u}} - \mathbf{D}^{\mathbf{u}}_n \right) - \int_{\mathcal{B}} \mathrm{d}V \frac{\rho}{\Delta t_{n+1} \Delta t_n} \mathbf{M} \left(\mathbf{D}^{\mathbf{u}}_n - \mathbf{D}^{\mathbf{u}}_{n-1} \right) \\ & + \int_{\mathcal{B}} \mathrm{d}V \left(\boldsymbol{\sigma}^h : \mathbf{S} \right)^T - \int_{\partial \mathcal{B}_J} \mathrm{d}S\,\mathbf{N}^T\,\bar{\mathbf{t}}^h, \end{aligned} \tag{3.45}$$

$$\mathbf{R}_d = \int_{\mathcal{B}} \mathrm{d}V \left(\frac{G_c}{l} + 2\mathcal{H}^h\right) \mathbf{M}\,\mathbf{D}^d + \int_{\mathcal{B}} \mathrm{d}V G_c l\, \mathbf{B}^T \mathbf{B}\,\mathbf{D}^d, \tag{3.46}$$

$$\mathbf{R}_{\sigma_h} = \int_{\mathcal{B}} \mathrm{d}V\, \mathbf{M}\,\mathbf{D}^{\sigma_h} - \int_{\mathcal{B}} \mathrm{d}V\, \frac{1}{3}\mathrm{tr}[\boldsymbol{\sigma}^h]\, \mathbf{N}^T. \tag{3.47}$$

The formal way of writing the expressions through ordered representations of scalars and tensors bares notational subtleties that shall be explained here. First, $\boldsymbol{\sigma}^h$ is the approximated Cauchy stress tensor, not the hydrostatic stress σ_h. Second, the product $\mathbf{N}^T\,\bar{\mathbf{t}}^h$ is computed in a component-wise way. Each component $\mathbf{N}^T$ is multiplied by the three-dimensional vector $\bar{\mathbf{t}}^h$ yielding an ordered representation of vectors that is consistent with the definition of $\mathbf{D}^{\mathbf{u}}$. Similar, $\left(\boldsymbol{\sigma}^h : \mathbf{S}\right)^T$ describes an array of column vectors $\boldsymbol{\sigma}^h : \mathbf{S}_j = (\sigma_{kl} S^j_{klm})$, where m is the index of a vector-component running from one to three, or x, y, z respectively[5]. The tangent matrices are then given by

$$\mathbf{K}_{\mathbf{uu}} = \int_{\mathcal{B}} \mathrm{d}V \frac{\rho}{\Delta t_{n+1}^2} \mathbf{M} + \int_{\mathcal{B}} \mathrm{d}V\, \left(\boldsymbol{\sigma}^h : \mathbf{S}\right)^T : \mathbf{S}, \tag{3.48}$$

$$\mathbf{K}_{dd} = \int_{\mathcal{B}} \mathrm{d}V \left(\frac{G_c}{l} + 2\mathcal{H}^h\right) \mathbf{M} + \int_{\mathcal{B}} \mathrm{d}V G_c l\, \mathbf{B}^T \mathbf{B}, \tag{3.49}$$

$$\mathbf{K}_{\sigma_h \sigma_h} = \int_{\mathcal{B}} \mathrm{d}V\, \mathbf{M}. \tag{3.50}$$

Here $\left(\boldsymbol{\sigma}^h : \mathbf{S}\right)^T : \mathbf{S}$ represents a $\mathcal{N} \times \mathcal{N}$ array of 3×3 tensors that are defined by[5] $\left(\left(\boldsymbol{\sigma}^h : \mathbf{S}\right)^T : \mathbf{S}\right)_{ij} = \mathbb{C}_{klmn} S^i_{mno} S^j_{klp}$, where $k, l, m, n, o, p \in \{0, 1, 2\}$ and $i, j \in \{0, 1, \ldots, \mathcal{N}\}$.

3.2 Implementation and Software

As mentioned in the introduction of this work, the application of the phase field method for crack growth in a coupled model with Li diffusion in storage particles bears several advantages. These are, for example, unconstrained crack paths, that are not determined in advance, or a

[5] Einstein's sum convention is used in this notation.

straight-forward integration into the existing model describing Li diffusion. However, the phase field method also presents numerical challenges that must not be underestimated.

In particular, the size of the systems[6] that are solved can easily reach levels that demand for an elaborate solution algorithm. The reason for this lies in the smoothing of the discontinuous crack over a certain length scale l. Only if this parameter is small in comparison to the relevant length scale of the problem, the phase field approximation is justified. Consider, for example, a two-dimensional specimen of square shape with side length $b = 10\,\mu\text{m}$, that has a triangular shaped notch on one side. Boundary conditions shall be applied in such a way that a crack starts to grow at the notch. In order for the phase field to recognize the notch as a distinct geometric feature, the length scale parameter l must be chosen such that it is significantly smaller than the notch size. Otherwise the local differences in the stress can not be distinguished correctly by the phase field. If the notch size is $a = 1\,\mu\text{m}$, a reasonable value for the length scale parameter is $l \approx 0.1\,\mu\text{m}$. To numerically resolve the phase field in a sufficiently precise way, the element size of the finite element discretization should be approximately[7] $h \approx l/2$ [109]. Assuming a uniform mesh of four-sided quadratic Lagrange elements (QUAD9), this requires around 160 000 nodes or, in our model, 800 000 degrees of freedom. Of course, these numbers are considerably exceeded in a three-dimensional computation.

The computational needs are further aggravated by the fact, that the system is time-dependent with two very different characteristic time scales, that it is nonlinear and that it comes with a bad conditioning of the elastic tangent matrix due to the decomposition into compressive and tensile stress when a crack opens.

Some of the strategies and tools, employed to overcome these hurdles, are outlined in the following sections.

3.2.1 deal.II

The typically large number of degrees of freedom, resulting from the phase field method for crack growth, is mainly tackled with two separate techniques. First, by using an adaptive mesh algorithm only regions

[6] In the sense of the number of degrees of freedom of the matrix equations.

[7] Alternative approaches of the phase field method for crack growth, which relax this rule of thumb [110, 152], are not further discussed in this work.

around a crack are resolved finely. In this way, the number of degrees of freedom is kept as small as possible without loosing significant numerical accuracy. Second, the still remarkably large matrix systems are assembled and solved in a parallel way spreading the computational tasks among multiple processors, as shown in FIG. 3.1.

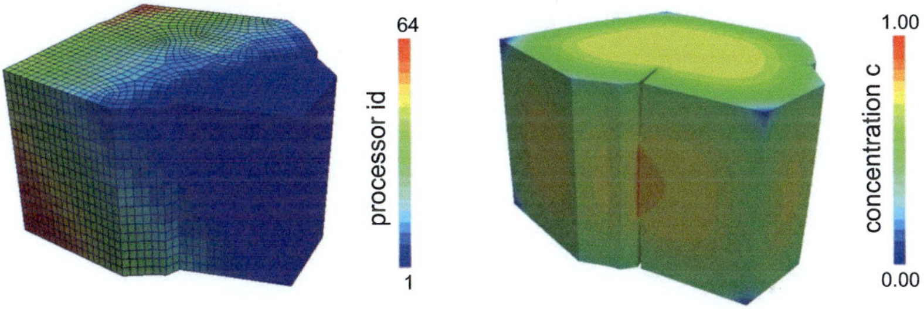

Figure 3.1.: The coupled problem of diffusion and fracture (right) is solved in a parallel way, distributing the numerical work among several processors (left). Each color on the left denotes one processor, so that only a small fraction of the overall element number has to be solved by one processor.

The combination of both methods creates several intricacies, such as how to distribute the computational load uniformly among the processors while the partitioned mesh is being refined and coarsened, or how to obtain required solutions of previous steps, that might have been computed on another processor using a different mesh.

To address these details in an efficient way, we implement the model in a C++ code using the finite element library deal.II[8] v8.1 [153, 154], an open source project maintained primarily at Texas A&M University, that originally emerged from work at the Numerical Methods Group at Universität Heidelberg.

The major scope of deal.II is to aid the development of finite element code for efficiently solving partial differential equations using adaptive meshes. It further supports implementation of the code in a parallel fashion, both on a single machine through Threading Build Blocks and across multiple computer nodes via the Message Passing Interface (MPI). To maintain work balance across the processors while adaptation of the mesh, deal.II can be interfaced to the software library p4est [155]. Its

[8] The abbreviation deal.II stands for being the successor of the so called Differential Equations Analysis Library.

task is the management of the mesh adaptation steps as a collection of adaptive octrees, i.e. in a tree data structure.

Apart from these essential properties, deal.II comes with a number of useful features. Two of them shall be mentioned briefly. First, it brings interfaces to the software projects Trilinos [156] and PETSc [157]. This enables us to choose from a large set of numerical solvers and preconditioners and pick those that are most suitable with respect to the properties of the corresponding matrix equations. We found that this has an immense impact on the stability and computation time of the simulations. Second, deal.II is set up in a way that allows to write the program in a dimension-independent way. We are therefore able to run simulations in both two and three dimensions with only minor changes in the code.

3.2.2 ParaView

ParaView [158] is an open-source data analysis and visualization application, that emerged to the largest part from a collaboration of Kitware, Sandia National Labs and CSimSoft. It was developed to analyze large datasets using distributed memory computing resources. It includes filters for different methods of visualizing ones data and readers for several data types. In particular, it is able to handle the combination of HDF5 files, a format for extremely large datasets, e.g. resulting from simulations on supercomputers, and XDMF metadata files, which are used to store the big amount of data generated during the simulations of this work.

3.2.3 Trelis

Trelis [159] is a commercial preprocessor for finite element analysis and computer fluid dynamics, that is based on the CUBIT geometry preparation and mesh generation software from Sandia National Laboratories. It is being distributed by CSimSoft. Among the myriad of available meshing software, we decided to pick Trelis for two reasons. On the one hand, the meshing algorithm is well developed for hexahedral elements. This is a constraint implied when using deal.II, which does not support other element types, such as, for example, tetrahedral ones. And on the other hand, it provides an interface, that allows to write Python scripts for the output of meshes into user-specific file formats. This interface showed great flexibility for the definition of specific boundary conditions and their exportation to the input file required by deal.II.

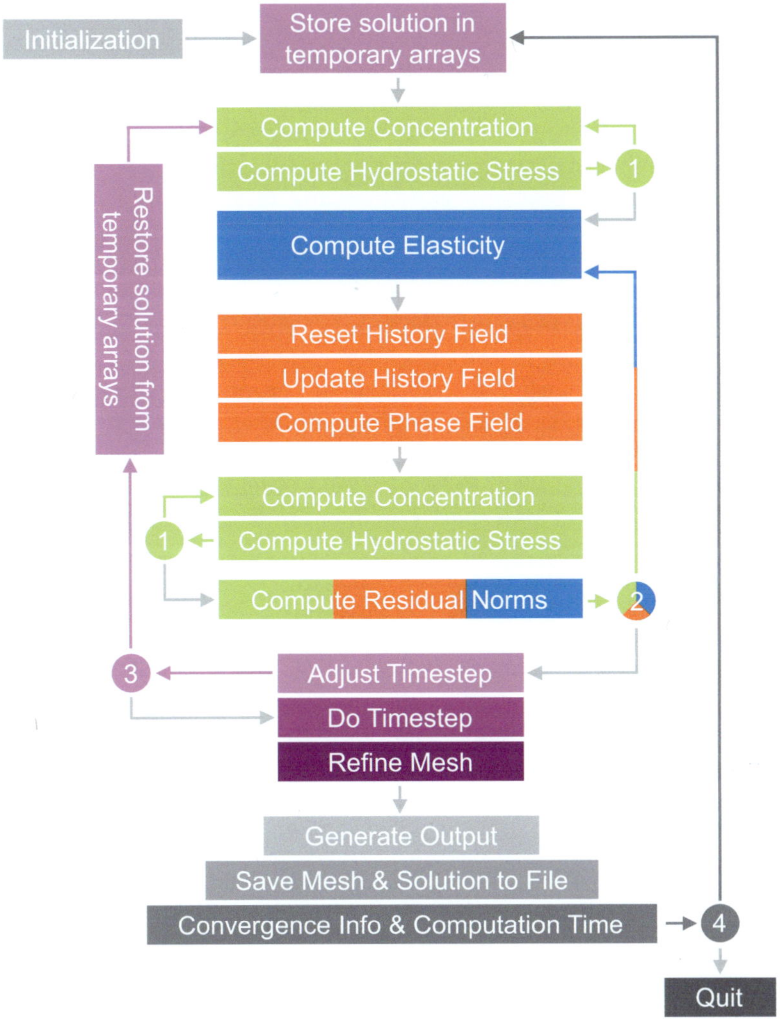

Figure 3.2.: Schematic structure of the algorithm. The purpose of the colored blocks is to solve the respective PDE's given in FIG. 2.2. Purple blocks refer to time or mesh adaptation steps and gray blocks denote further auxiliary functions and operations.

3.3 Structure and Details of the Implementation

The basic structure of the `C++` code is presented in the diagram of FIG. 3.2. The blocks shown in the diagram are marked with different colors according to their task in the computer algorithm. The purpose of the green, blue and orange blocks is to solve the respective part of the boundary value problem, corresponding to the partial differential equations given in FIG. 2.2. Gray blocks refer to auxiliary functions and operations, such as generating data for visualization or analysis of convergence behavior and computation time. Functions related to time and mesh adaptation are printed in purple. The numbers in circles represent `if`-cases in the computer code that decide whether the algorithm proceeds with the next step or repeats the respective step in a loop. The criteria of the `if`-cases will be explained in the corresponding sections below. Colored and dark gray arrows refer to steps that are executed in a loop, light gray arrows denote simple proceeding from one block to the next one. The diagram is read from the top left to the bottom right.

The first steps in the computer code are summarized by the term *initialization* in FIG. 2.2. This includes setting up the MPI-communication between different processors and reading parameters from an input file, e.g. material and numerical parameters, geometry and mesh information, boundary conditions, and more. According to the information given in the input file, the mesh is either constructed within deal.II or loaded from a mesh file. Special types of parallelized arrays[9] are then set up for later storage of residual vectors and tangent matrices. Thereafter the initial conditions are either defined from values given in the input file or restored from previous solutions in a saved file. This is typically succeeded by a number of mesh refinement steps, either to provide a mesh that is fine enough with respect to the criteria discussed in section 3.3.5 or to reconstruct the final mesh of the previous simulation.

After completion of the initialization step, the actual solution process begins. It is built up by essentially three nested loops. The outermost loop, depicted by the dark gray arrows starting and ending at the circle

[9] These objects are provided by deal.II and hold additional information, e.g. about eventual constraints such as Dirichlet boundary conditions and hanging-node constraints, or sparsity properties of the tangent matrices. In addition, versions optimized for the exchange between deal.II and Trilinos or PETSc can be selected.

with number 4 iterates through the number of time steps. It stops when the maximum time defined in the input file is exceeded. For a simple transient linear problem with constant time steps, this would be the only loop required to obtain the solution at the respective time steps. Within the iterations, the current solution is computed, the mesh is refined and coarsened, output for later visualization is generated, information on convergence behavior and computation time is printed and the solution, as well as the mesh, is saved in a file. In order to save computation time, some of these tasks are executed only every x-th time step. In addition to these steps, the time-adaptive method adjusts the time step after the current time step has been computed.

As described in section 3.3.4, the time adaptation algorithm is implemented in such a way that the current time step is re-calculated when the used time step size turns out to be too large. The solution of the previous time step is therefore stored in temporary arrays before solving for the new solution and is restored in the case of a decrease of the time step, as depicted by the upper two purple blocks in FIG. 2.2. In this case, the algorithm runs a limited number of time adaptation loops, as illustrated by the purple arrows. The decision on whether the time step is adjusted, is marked by the purple circle with number 3 and follows the criteria described in the scheme presented in section 3.3.4.

The purple blocks of the time adaptation scheme frame the central part of the computer code, where the partial differential equations of the boundary value problem are solved. As mentioned earlier, the solution algorithm is built in a staggered fashion using a mixed formulation, that takes the hydrostatic stress as an additional variable. As a consequence, the equations for the balance of Li and linear momentum, the definition of the hydrostatic stress and the phase field equation are solved separately while holding the other variables fixed. Using a mixed formulation and splitting the equation for balance of Li into two equations does, in principal, not imply any kind of approximation. Solution of both equations in a coupled form by an exact method yields the same results as the solution of the equation for balance of Li expressed without the additional definition of the hydrostatic stress. However, the treatment of the two equations in a staggered fashion, i.e. one after the other, introduces an error due to their mutual dependence. In order to reduce this error, both equations are always implemented in an internal loop, as shown by the green arrows. Only when their combined residual norm falls below a certain threshold, the solution of the diffusion part is accepted and the algorithm steps on. This criterion is marked by the green circles with number 1. The number of these iterations is exploited in the time adaptation scheme, described in section 3.3.4, since it yields an estimate on the dynamics of the diffusion problem at the current time step, e.g. whether the concentration is changing rapidly with large spatial gradients or flowing more steadily. In addition, since the equation for balance of Li and, even more, for the hydrostatic stress are comparatively fast to assemble and solve, it is also advantageous in terms of the total computation time to solve them more often than, in particular, the equation for balance of linear momentum.

After the first solution of the concentration and hydrostatic stress fields is obtained (in the green blocks of 2.2), the different physical parts of the problem are solved in a number of iterations, as depicted by the tricolor arrow. Thereby, the equation for balance of linear momentum (blue) is computed first. Thereafter the history and phase field (orange) are calculated. The former step is divided into two blocks, a subtlety that is described in section 3.3.3. In the final step, the diffusion part of the boundary value problem is solved again. The combined residual norm is calculated as

$$R_{\text{total}} := \sqrt{|\mathbf{R}_c(\mathbf{D}^c)|^2 + |\mathbf{R}_{\sigma_h}(\mathbf{D}^{\sigma_h})|^2 + |\mathbf{R}_{\boldsymbol{u}}(\mathbf{D}^{\boldsymbol{u}})|^2 + |\mathbf{R}_d(\mathbf{D}^d)|^2} \tag{3.51}$$

and compared with a threshold value to determine whether another iteration is required. This criterion is presented by the tricolor circle with number 2. After either reaching the desired precision or the maximum number of iterations, the algorithm steps on. Depending on the time adjustment, it either proceeds towards the next time step or re-calculates the current one.

Some of the functions and blocks presented above bear intricate details and shall therefore be discussed in more detail in the following sections.

3.3.1 Linear Blocks

The two linear equations in our model determine the hydrostatic stress σ_h and the phase field d by means of (2.17) and (2.29) respectively. Due to the linearity of the equations, it is sufficient to assemble the corresponding tangents and residuals once and then solve them with respect to the desired precision. In fact, the matrix resulting from the discretization of the definition of the hydrostatic stress is constant and only has to be built from scratch when the mesh is refined or coarsened. It is commonly referred to as mass matrix at unit density. Being both symmetric and positive definite, this matrix is particularly easy to solve. An ordinary conjugate gradient (CG) solver with a symmetric successive over-relaxation (SSOR) preconditioner turned out to be a good choice to approach this equation. The equation describing the evolution of the phase field is a Helmholtz type equation and shares the same properties. The corresponding tangent matrix is symmetric and positive definite. However, due to the large local differences in the history field, a more conservative selection of the solver showed to be advantageous. We therefore pick the generalized minimal residual method (GMRES) with the algebraic multigrid preconditioner (AMG) from the Trilinos package.

3.3.2 Nonlinear Blocks

The nonlinear equations in our model describe the balance of Li content and the balance of linear momentum. Different numerical approaches were used to solve the equations and shall be presented here. The nonlinearity in the **balance of linear momentum** stems from the spectral decomposition of the strain tensor into tensile and compressive parts. As soon as the displacement field is updated by the new solution, the eigenvalues of the strain tensor change and the corresponding stiffness tensor has to be adapted. In order to find an optimal convergence of the non-linear iterations in the sense of rate and stability, we implemented a backtracking line search method as explained in [160]. Instead of just adding the computed increment from equation (3.42) to the old solution, we first determine an optimal step length α to give

$$\mathbf{D}^{\mathbf{u}}_{\text{new}} = \mathbf{D}^{\mathbf{u}}_{\text{old}} + \alpha\,\Delta\mathbf{D}^{\mathbf{u}}. \tag{3.52}$$

In principal, the convergence of the Newton-Raphson method is best, i.e. quadratic, close to the solution, when $\alpha = 1$. However, when the starting point is not close enough to the solution, setting α to unity can lead to over-shooting, which may slow down convergence or even result in divergence. To avoid this behavior, we imply the so called *sufficient decrease* or *Armijo condition*

$$|\mathbf{R}_{\mathbf{u}}(\mathbf{D}^{\mathbf{u}}_{\text{old}} + \alpha\,\Delta\mathbf{D}^{\mathbf{u}})| \leq |\mathbf{R}_{\mathbf{u}}(\mathbf{D}^{\mathbf{u}}_{\text{old}})| + c_1\alpha\,|\mathbf{K}_{\mathbf{uu}}\Delta\mathbf{D}^{\mathbf{u}}|, \tag{3.53}$$

where $c_1 \in (0,1)$ and the Euclidian norm $|\mathbf{x}| = \sqrt{x_1^2 + x_2^2 + \cdots + x_n^2}$. The condition only ensures, that the step length is not too large and may be extended by further requirements[10] to determine also a lower boundary for α. We go without an additional condition and use a backtracking approach to find the largest step length, that still fulfills equation (3.53). The corresponding algorithm looks schematically as follows:

1. Set q, $c_1 \in (0,1)$ and $\alpha = 1$.
2. Solve $\mathbf{R_u} + \mathbf{K_{uu}}\Delta\mathbf{D^u} = \mathbf{0}$ (3.42) to obtain $\Delta\mathbf{D^u}$.
3. Compute the old residual $\mathbf{R_u}(\mathbf{D^u_{\text{old}}})$ and direc. derivative $\mathbf{K_{uu}}\Delta\mathbf{D^u}$.
4. Compute the new residual $\mathbf{R_u}(\mathbf{D^u_{\text{old}}} + \alpha\,\Delta\mathbf{D^u})$.
5. Check the *sufficient decrease condition* (3.53).
   a) If the condition is fulfilled, update the old solution to get $\mathbf{D^u_{\text{new}}} = \mathbf{D^u_{\text{old}}} + \alpha\,\Delta\mathbf{D^u}$.
   b) If the condition is not fulfilled, decrease the step size by $\alpha \leftarrow q\,\alpha$ and go back to 4.
6. If for the new residual norm $|\mathbf{R_u}(\mathbf{D^u_{\text{new}}})| >$ threshold, assemble the new residual and tangent matrix and go back to 1. Else, quit.

The second step in the above scheme is the computationally most expensive part in the overall algorithm. The elastic equation is of vector-type and has dim$\times\mathcal{N}$ degrees of freedom, where $\mathcal{N}$ is the number of nodes and dim is the number of physical dimensions. The impact of the number of degrees of freedom on the computation time is further enhanced since the solution time of iterative solvers[11] typically scales worse than linear with the number of degrees of freedom. In addition, the tangent matrix can become badly conditioned due to the decomposition into tensile and compressive parts. The most effective preconditioner was found to be the algebraic multigrid preconditioner (AMG) from the Trilinos package. Since the matrix is both positive definite and symmetric, we combined it with a conjugate gradient (CG) solver. To reduce the number of calls

[10] See [160] for the so called *curvature condition*.

[11] The computation time of iterative solvers typically scales between linear and quadratic with the number of degrees of freedom, whereas that of direct solvers scales between quadratic and cubic. Due to this behavior, the larger memory requirements and drawbacks in terms of the parallel performance, we refrain from using direct solvers.

of the computationally expensive solver, we further integrate a prediction step into the first iteration. This step generates both a linear and a quadratic extrapolation of the displacements using the last time steps. By a small number of bi-sections it then finds the linear combination of both extrapolations, which minimizes the residual norm $|\mathbf{R_u}|$. This prediction function significantly helps to lower the number of nonlinear iterations and requires only negligible additional computation time.

The nonlinear part in the **balance of Li content** originates from the mechanically driven term in the Li flux (2.20). Due to the non-constant mobility, this term is quadratic in the concentration. Further, the mixed formulation, that takes the hydrostatic stress as a separate variable, comes with an additional numerical subtlety. In general, it would be most common to solve the conservation of Li and the hydrostatic stress as one system. However, in our case it turned out, that this approach does not perform well in terms of stability. Instead, we work with a staggered scheme, as shown by the green blocks and arrows in FIG. 2.2, of the following form:

1. Set $\beta_0 \in (0,1)$, $n_\beta > 1$ and $j = 1$.
2. Solve $\mathbf{R}_c + \mathbf{K}_{cc}\Delta\mathbf{D}^c = \mathbf{0}$ (3.39) to obtain $\Delta\mathbf{D}^c$ and $\mathbf{R}_{\sigma_h} + \mathbf{K}_{\sigma_h\sigma_h}\Delta\mathbf{D}^{\sigma_h} = \mathbf{0}$ (3.44) to get $\Delta\mathbf{D}^{\sigma_h}$.
3. If $j < n_\beta$, set the step length $\beta = \beta_0 + j(1-\beta_0)/n_\beta$. Else $\beta = 1$.
4. Update the solution to receive the new concentration and hydrostatic stress $\mathbf{D}^c_{\text{new}} = \mathbf{D}^c_{\text{old}} + \beta\,\Delta\mathbf{D}^c$ and $\mathbf{D}^{\sigma_h}_{\text{new}} = \mathbf{D}^{\sigma_h}_{\text{old}} + \beta\,\Delta\mathbf{D}^{\sigma_h}$.
5. Compute the combined residual norm
$$R_{\text{diff}}(\beta) := \sqrt{|\mathbf{R}_c(\mathbf{D}^c_{\text{new}})|^2 + |\mathbf{R}_{\sigma_h}(\mathbf{D}^{\sigma_h}_{\text{new}})|^2}.$$
6. If for the combined residual norm $R_{\text{diff}}(\beta) >$ threshold, assemble the new residual vectors and tangent matrices, increase the loop variable $j \leftarrow j+1$ and go back to 2. Else, quit.

The presented scheme sets the step length β in a predefined way without checking any optimality condition. If β_0 is small, this might result in an unnecessarily large number of iterations for finding the current solution. However, the approach outperformed more sophisticated techniques in terms of both stability and computation time. There are several reasons for this. On the one hand, no directional derivatives and only a few number of residuals have to be computed. On the other hand, the increasing step length β is similar to an approach, where the diffusion part of the boundary value problem is solved with smaller time steps between the elastic block. In combination with the time adaptation scheme, presented in section (3.3.4), this turned out to be a very effective strategy.

3.3.3 Updates of the History Field

The solution blocks of the equations describing the balance of linear momentum, abbreviated by the term *Elasticity* in the diagram, and the evolution of the phase field are combined into the tricolor loop of FIG. 3.2, which assures that the total residual norm converges below a certain threshold. Inside this loop the history field, which represents the driving force for the phase field, is only updated temporarily. This is achieved with the following scheme:

1. Assemble and solve the balance of linear momentum as described in section 3.3.2 to obtain the current displacement solution $\mathbf{D}^{\mathbf{u}}$. *See first blue block in FIG. 3.2.*

2. Store the previous history field $\mathbf{D}^{\mathcal{H}}$ in the temporary field $\mathbf{D}^{\mathcal{H}}_{\text{tmp}}$. *See first orange block in FIG. 3.2.*

3. Update $\mathbf{D}^{\mathcal{H}}$ as node-wise maximum

$$(\mathbf{D}^{\mathcal{H}})_j = \max[(\mathbf{D}^{\mathcal{H}}_{\text{tmp}})_j, (\mathbf{D}^{\psi^+_{\text{mech}}})_j],$$

   where $\mathbf{D}^{\psi^+_{\text{mech}}}$ are the nodal values of the elastic, tensile energy ψ^+_{mech}, obtained via extrapolation from Gaussian integration points. *See the second orange block in FIG. 3.2.*

4. Assemble and solve the equation describing the evolution of the phase field to obtain the current solution of the phase field $\mathbf{D}^d$. *See the third orange block in FIG. 3.2.*

5. Solve the diffusion part of the BVP as described in section 3.3.2 to obtain the current solution of the equation describing balance of Li content $\mathbf{D}^c$ and the definition of the hydrostatic stress $\mathbf{D}^{\sigma_h}$. *See the third and fourth green blocks in FIG. 3.2.*

6. Compute the total residual norm R_{total} via equation (3.51).
   See the tricolor block in FIG. 3.2.

7. If the residual norm $R_{\text{total}} >$ threshold, go back to 2. Else, quit.
   See the tricolor circle with number 2 in FIG. 3.2.

As a result of this strategy, the history field is always a result of fully converged solutions of the displacement and phase field. The approach assures that values from non-converged solutions do not affect the evolution of the phase field and therefore the predictions on crack growth.

3.3.4 Time Adaptivity

The time adaptation scheme, implemented in our code, intends to find the maximum time step size, which is still small enough to describe both the process of Li diffusion and crack propagation in a satisfactorily accurate way. It turned out, that a good indicator with respect to this directive is given by the number of iterations to reach a certain threshold for the combined residual norms R_{diff} and R_{total}. The initiation and subsequent growth of a crack may happen in a rather drastic event. Consequently, the time-adaptive method must be able to react fast enough to such incidents. To meet this requirement, we formulate the algorithm in such a way that it allows to go one time step back and re-calculate the current time step, if the time step size turns out to be too large. The scheme thus reads like the following:

1. Set factors to adapt time step $k_{\downarrow}^{\text{total}} \in (0,1)$, $k_{\downarrow}^{\text{diff}} \in (0,1)$ and $k_{\uparrow} > 1$ and maximum and minimum number of iterations $n_{\text{max}}^{\text{total}}$, $n_{\text{max}}^{\text{diff}}$ and $n_{\text{min}}^{\text{total}}$, $n_{\text{min}}^{\text{diff}}$ for solving the full BVP and its diffusion part.

2. Store the current solution of the displacements $\mathbf{D^u}$, concentration $\mathbf{D}^c$, hydrostatic stress $\mathbf{D}^{\sigma_h}$, as well as the history field $\mathbf{D}^{\mathcal{H}}$ in temporary vectors $\mathbf{D}_{\text{tmp}}^{\mathbf{u}}$, $\mathbf{D}_{\text{tmp}}^{c}$, $\mathbf{D}_{\text{tmp}}^{\sigma_h}$, $\mathbf{D}_{\text{tmp}}^{d}$ and $\mathbf{D}_{\text{tmp}}^{\mathcal{H}}$.
   See the first purple block in FIG. 3.2.

3. Compute the blocks related to the concentration, hydrostatic stress, elasticity and phase field.
   See the green, blue and orange blocks in FIG. 3.2.

4. Check the number of iterations n^{total} and n^{diff} to pass the desired thresholds with residual norms R_{total} and R_{diff}
   See the second purple block and purple circle in FIG. 3.2.

a) If $n^{\text{total}} > n^{\text{total}}_{\text{max}}$ or $n^{\text{diff}} > n^{\text{diff}}_{\text{max}}$, go one time step back by setting $t \leftarrow t - \Delta t$ and restore the temporary solutions $\mathbf{D}^{\mathbf{u}} \leftarrow \mathbf{D}^{\mathbf{u}}_{\text{tmp}}$, $\mathbf{D}^{c} \leftarrow \mathbf{D}^{c}_{\text{tmp}}$, $\mathbf{D}^{\mathcal{H}} \leftarrow \mathbf{D}^{\mathcal{H}}_{\text{tmp}}$ and $\mathbf{D}^{d} \leftarrow \mathbf{D}^{d}_{\text{tmp}}$.

   i. If $n^{\text{total}} > n^{\text{total}}_{\text{max}}$, decrease time step via $\Delta t \leftarrow k^{\text{total}}_{\downarrow} \Delta t$.

   ii. Else if $n^{\text{diff}} > n^{\text{diff}}_{\text{max}}$, decrease time step via $\Delta t \leftarrow k^{\text{diff}}_{\downarrow} \Delta t$.

   Go back to 3.
   See the vertical purple block in FIG. 3.2.

b) Else if $n^{\text{total}} < n^{\text{total}}_{\text{min}}$ or $n^{\text{diff}} < n^{\text{diff}}_{\text{min}}$, increase the time step via $\Delta t \leftarrow k_{\uparrow} \Delta t$.

The factor for the decrease of the time step due to the total iterations $k^{\text{total}}_{\downarrow}$, which is dominantly influenced by the residual norms of the solution of the displacements and the phase field when a crack grows, is typically set smaller than the one for the diffusion part $k^{\text{diff}}_{\downarrow}$. This way, the code is both able to find a good time step for the relatively smooth diffusion process and may adjust quickly to the appearance of crack growth. The necessity for this distinction becomes obvious by having a look at the typical time step lengths attained during different dominant physical effects. A sole diffusion process, for example, typically requires a time step[12] of around $\Delta t/\tau \approx 10^{-2}$, while during unstable crack growth, a time step of around $\Delta t/\tau \approx 10^{-13}$ is adequate. It thus requires a drastic drop of the time step to resolve correctly the growth of a crack from its first onset on.

Furthermore, the large difference in the order of magnitude of Δt demands for a careful handling of numerical errors. The changes in the concentration that appear, for example, during a time step as short as $\Delta t \approx 10^{-13}$ may be very small in comparison to the given numerical precision. Hence, small errors might be introduced. Due to the large number of required time steps, these errors may sum up over time. To counteract this phenomenon, we temporarily disable the diffusion block in the solution algorithm in cases where $\Delta t < \Delta t_{\text{min}}$. Instead, we sum up the time steps $\Delta t_{\text{diff}} = \sum_i \Delta t_i$ until $\Delta t_{\text{diff}} \geq \Delta t_{\text{min}}$ and then solve the diffusion equation with this time increment.

[12] The characteristic time scale $\tau = r_0^2/D_0$ involved in the Li diffusion problem is defined through the diffusion coefficient D_0 and the characteristic length scale, which is set to $r_0 = 1\,\mu\text{m}$ in this work.

3.3.5 Mesh Adaptivity

The indispensability of a mesh adaptation method was outlined in the introduction of section 3.2. The basic idea of this method is to refine the mesh only in regions where a high spatial resolution is required. As a result, the number of elements or degrees of freedom can be reduced substantially.

Using the phase field method for crack growth, the possible savings in computation time acquired with a mesh-adaptive method are even enhanced due to the correlation between element size h and length scale parameter l. Imagine, for example, a two-dimensional geometry with surface area A. Assuming approximately equally large elements with area h^2, the number of elements is given by $N_{\text{elem}} \approx A/h^2$. As mentioned earlier, the element size should be at maximum equal to half of the value of the length scale parameter, i.e. $h \approx l/2$ [109]. This means, that the number of elements goes with $N_{\text{elem}} \propto A/l^2$. Imagine now that not the whole geometry but only a crack in its interior is covered with such a fine mesh. The crack shall have the length a and be described by a smooth phase field $d = d(\mathbf{x})$. The area that needs to be covered by a fine mesh is determined by the region where d is notably above zero[13] since the spatial variation of d is negligible in other parts of the geometry. The width of this area shall be denoted by b. Assuming that the mesh is considerably finer at the crack than at the rest of the geometry, the greatest part of elements is located around the crack. The total number of elements is therefore given by $N_{\text{elem}} \approx ab/h^2$ or $N_{\text{elem}} \propto ab/l^2$. However, the width of the area is only a numerical parameter, which becomes smaller with a decreasing length scale parameter, namely by $b \propto l$. The number of elements therefore scales linearly with the inverse of the length scale parameter[14], i.e. $N_{\text{elem}} \propto a/l$. This is in contrast to the quadratic dependence in the case of a uniformly refined mesh. Consequently, using the same total number of elements, much smaller values for l can be achieved with an adaptive mesh refinement method.

The obligation of using element sizes, determined by the length scale parameter, lies in the computation of the crack volume. If the shape of the phase field is represented with only insufficient accuracy, the fracture energy cannot be calculated in a satisfying way and the phase field may not evolve the same as it would with a finer mesh. In addition, geomet-

[13] For example, in the studies of chapters 4 to 9, a value of $d = 0.95$ was defined to distinguish between cracked and intact material.

[14] A similar argument holds for three-dimensional geometries.

rical details that could act as stress concentrators may not be resolved accurately enough. This can lead to an underestimation of the elastic energy, i.e. the driving force for the phase field, in those points. In order to obtain a mesh-independent phase field evolution, it is therefore mandatory to refine the mesh at the relevant positions before a crack even starts to grow.

To satisfy this requirement, we set up an algorithm, that uses the local elastic energy from tension ψ^+ as a trigger value to either refine or coarsen the mesh. The choice of ψ^+ is most suitable since it defines the driving force of the phase field through the history field $\mathcal{H}$. Hence, before crack growth is initiated, ψ^+ is expected to increase locally. This increase is used as a criterion to trigger mesh refinement before a crack begins to propagate. The tensile elastic energy ψ^+ is therefore compared with a threshold value ψ_{thresh}, so that an element is refined if $\psi^+ > \psi_{\text{thresh}}$. As mentioned above, insufficiently accurate resolution of stress concentrators in an initially coarse mesh may lead to a numerical underestimation of the local tensile energy ψ^+. To take into account this effect, ψ_{thresh} should therefore be comparatively small in regions where the mesh is coarse. On the other hand, it is computationally desirable to generate a very fine mesh only in regions of maximum tensile elastic energy, i.e. in regions where a crack is expected to grow. Hence, in a finely meshed area, ψ_{thresh} should be comparatively large so that further refinement only takes place around the locations of highest values of ψ^+. To manage both requirements, we assume that the initial mesh is built up by approximately equally sized elements and define the threshold as a function $\psi_{\text{thresh}} = \psi_{\text{thresh}}(n)$, that depends on the number of refinements n of the element in question by

$$\psi_{\text{thresh}}(n) = (\psi_{\text{up}} - \psi_{\text{low}}) \left(\frac{n - n_{\text{min}}}{n_{\text{max}} - n_{\text{min}}} \right)^r + \psi_{\text{low}}. \tag{3.54}$$

It is small for low numbers of refinements and increases with n. Here, ψ_{up} is the upper threshold for which all elements, that have been refined n_{max} times or less, are further refined. Likewise, ψ_{low} is the lower boundary at which elements are refined, that were only refined n_{min} times or less. Together with the initial element size, n_{max} and n_{min} determine the final smallest and largest element size obtained through this refinement criterion. The exponent r can be adjusted to determine the shape of the refinement region, e.g. to obtain large areas with only weak refinement and a small center with highly refined elements.

Elements are coarsened if $n > n_{\mathrm{min}}$ and if ψ_+ is lower than the value of the threshold function for a refinement number of $n-1$. This offset is introduced in order to avoid alternating refinement and coarsening when $\psi_{\mathrm{thresh}}(n) < \psi_+ < \psi_{\mathrm{thresh}}(n+1)$. The principle of the refinement and coarsening scheme is illustrated in FIG. 3.3.

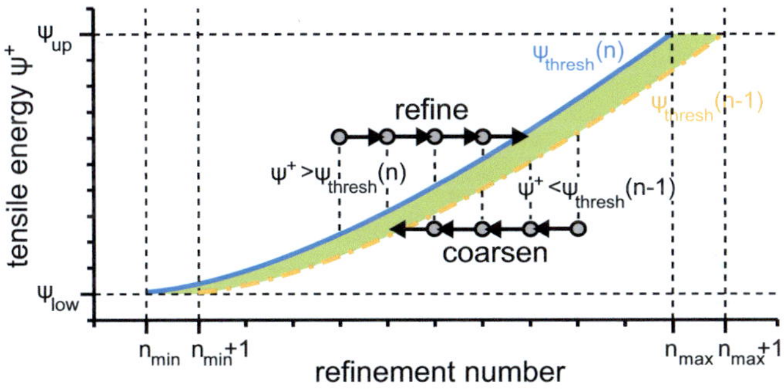

Figure 3.3.: Threshold function for mesh adaptation. Above the blue graph, the tensile energy exceeds the threshold and the element is refined. To avoid jumping back and forth coarsening is only allowed below the orange graph, where $\psi_+ < \psi_{\mathrm{thresh}}(n+1)$. In the green area, elements are neither refined nor coarsened.

The refinement parameter n_{max} must be chosen in such a way, that the most refined element size h_{min} is equal or less than half the length scale parameter l. At the same time, n_{min} determines the ratio of smallest and largest element size $h_{\mathrm{ratio}} = h_{\mathrm{max}}/h_{\mathrm{min}}$ and should be defined appropriately. If h_{ratio} is too small, the total number of elements will become unnecessarily high, if h_{ratio} is too large, the greater difference in element sizes might lead to stronger numerical errors. Since h_{min} and h_{ratio} are more general in terms of the initial mesh and more intuitively to set, we substitute the refinement numbers via $h_{\mathrm{min}} = h_0/2^{n_{\mathrm{max}}}$, $h_{\mathrm{max}} = h_0/2^{n_{\mathrm{min}}}$ and $h = h_0/2^n$, where h_0 is the initial element size. Inverting these relations and inserting them into equation (3.54) yields

$$\tilde{\psi}_{\mathrm{thresh}}(h) = (\psi_{\mathrm{up}} - \psi_{\mathrm{low}}) \left(1 + \frac{\ln(h_{\mathrm{min}})}{\ln(h_{\mathrm{ratio}})} - \frac{\ln(h)}{\ln(h_{\mathrm{ratio}})}\right)^r + \psi_{\mathrm{low}}, \quad (3.55)$$

which becomes independent from the initial element size. The so defined threshold function is thus also suitable for an initially non-uniform mesh.

The threshold function provides an energy based criterion for mesh refinement or coarsening. However, in situations, where a crack has developed and the tensile energy decreases again, the threshold function is not sufficient to guarantee, that the mesh remains fine enough. For this reason, a second measure is introduced.

The reason for the usage of element sizes equal or close to half the length scale parameter lies in a correct resolution of the gradient of the phase field. As a guideline, we take the one-dimensional solution (2.22) and derive the highest value of the phase field gradient

$$d'_{\max} = \lim_{|x|\to 0} |d'(x)| = \lim_{|x|\to 0} \frac{1}{l} e^{-|x|/l} = \frac{1}{l}. \tag{3.56}$$

As mentioned above, the element size should be at maximum equal to half the value of the length scale parameter, i.e. $h_{\min} = k\,l$ with $k \leq 1/2$. With this element size, the product of the highest value of the phase field gradient and the element size yields $d'_{\max} h_{\min} = k$, so that k determines the maximum change in the phase field over one element. Assuming that this value should not be exceeded in any part of the mesh, the upper boundary for the local element size is given by the relation $d'h \leq k$.

We generalize this argument to two and three dimensions and impose a second criterion, that triggers refinement if the product of the norm of the gradient of the phase field $|\nabla d|$ and the element size h is larger than a defined constant $|\nabla d|\,h > k_1$. Typically, k_1 is set slightly below one half. In the same way, coarsening is allowed if $|\nabla d|\,h < k_2$, where $k_2 \leq k_1/2$. The full refinement scheme, executed for each element, then looks like the following:

1. Loop over all quadrature points $\mathbf{q}_i \in \mathcal{Q}$ and find the maximum energy and gradient norm $\max\limits_{\mathbf{q}_i \in \mathcal{Q}} \psi^+$ and $\max\limits_{\mathbf{q}_i \in \mathcal{Q}} |\nabla d|$.

2. Compute the element size h as the maximum diameter divided by the root of the number of dimensions.

3. Compute the threshold energy for refinement and coarsening
   a) If $h \leq h_{\max}$, $\psi^{\text{refine}}_{\text{thresh}} = \tilde{\psi}_{\text{thresh}}(h)$ and $\psi^{\text{coarsen}}_{\text{thresh}} = \tilde{\psi}_{\text{thresh}}(h/2)$.
   b) Else, $\psi^{\text{refine}}_{\text{thresh}} = \tilde{\psi}_{\text{thresh}}(h_{\max})$ and $\psi^{\text{coarsen}}_{\text{thresh}} = \tilde{\psi}_{\text{thresh}}(h_{\max}/2)$.

4. For every element, set refinement and coarsening flags according to the following rules:

a) If $\max_{\mathbf{q}_i \in \mathcal{Q}} \psi^+ > \psi_{\text{thresh}}^{\text{refine}}$ and $h \geq h_{\min}$, flag for refinement.

b) Else if $\max_{\mathbf{q}_i \in \mathcal{Q}} |\nabla d| > k_1/h$, flag for refinement.

c) Else if $\max_{\mathbf{q}_i \in \mathcal{Q}} \psi^+ < \psi_{\text{thresh}}^{\text{coarsen}}$ and $\max_{\mathbf{q}_i \in \mathcal{Q}} |\nabla d| < k_2/h$ and $h \leq h_{\max}$, flag for coarsening.

d) Else, clear any refinement or coarsening flag.

After having run through all elements of the mesh, the algorithm may flag further elements to generate a smoother mesh and to avoid unfavorable constellations, e.g. isolated groupings of refined elements, that may lead to a degradation of the numerical approximation. The algorithms for this step are provided by deal.II.

A working example of the mesh refinement and coarsening algorithm is presented in FIG. 3.4. It is taken from chapter 4 and shows an edge notch specimen with an initial crack introduced as a free surface located along the left half of the bottom edge of the square. As the tensile load on the top surface is increased from a.i) to a.v), the phase field develops from $d = 0$ (blue) to $d = 1$ (red). It is observable that mesh refinement takes place before the phase field increases. As stress is released in certain regions due to the growth of the crack, the mesh is coarsened again, as can be seen by comparison of a.iv) and a.v). Nevertheless, the mesh remains fine enough to resolve the gradient of the phase field. This is shown in FIG. 3.4 b), where the product of the maximum phase field gradient and the local element size is plotted for the same time step as in FIG. 3.4 a.v). The product ranges from 0 (blue) to 0.44 (red). Hence, it is below the critical value of 0.5 in all regions of the specimen.

3.3.6 Performance and Scalability

The parallel scalability of the computer code is investigated in two and three-dimensional simulations. The corresponding physical problems are explained in detail in section 6.1 and 6.2. The scalability is tested with respect to strong scaling, which means that while the size of the problem is kept fixed, the number of processors is varied.

The first example runs with a mesh of 49 404 elements or 992 925 degrees of freedom in two dimensions. The required time to finish the first 20 time steps of the simulation is shown in FIG. 3.5. In order to obtain a better estimate of the long term scaling, i.e. in cases where the simulation runs to its end, we subtract the time necessary for initialization, for example, due to mesh construction or set-up of initial conditions, from the total time.

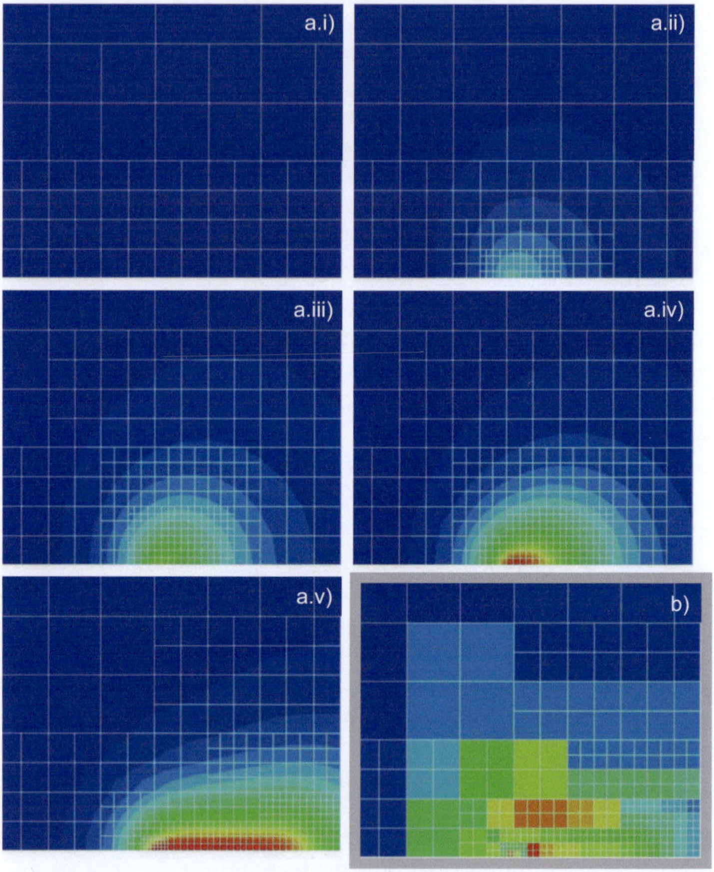

Figure 3.4.: Illustration of the adaptive mesh refinement and coarsening with a.i)–a.v) showing the phase field d with values ranging from $d = 0$ (blue) to $d = 1$ (red) and b) the product of maximum phase field gradient and element size with values ranging from 0 (blue) to 0.44 (red).

For small number of processors n_{cpu}, the scaling behavior of the program is almost directly linear, i.e. the computation time follows a simple relation of the form $t = t(n_{\text{cpu}} = 1)/n_{\text{cpu}}$. For $n \geq 8$, the scaling becomes sublinear, but is still satisfying, as shown in FIG. 3.5. The time saved using, for example, 32 instead of one processor, is still around a factor of 21. The scaling factor is thus approximately $n_{\text{cpu}} \times t(n_{\text{cpu}} = 32)/t(n_{\text{cpu}} = 1) \approx 2/3$. For higher numbers of processors, the scaling becomes significantly worse. The trade-off between distributing computational tasks and communication between the processors is getting less favorable for the system size of this example.

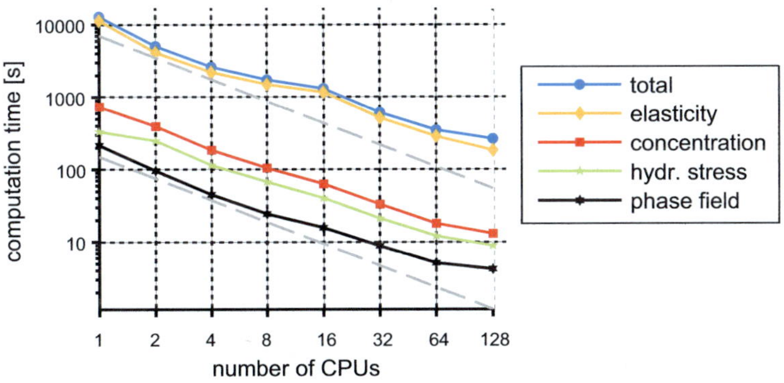

Figure 3.5.: Computation time for the first 20 time steps of the Li insertion example in 2D with approximately 1 million degrees of freedom. The dashed gray lines illustrate a perfectly linear scaling behavior, i.e. with $t = t_0/n_{\text{cpu}}$.

For systems with larger numbers of degrees of freedom the time for communication between the processors should become less significant in comparison to the reduction of computation time due to the distribution of computational tasks. This would result in a scaling behavior that is more closely to a linear one even for higher numbers of processors. In order to test this, we run the same example with a mesh that was additionally refined twice. The number of elements thereby increases to 786 441 with around 15 million degrees of freedom. Simulations of this size cannot be run in a reasonable amount of time on a single processor. However, the linear scaling behavior can not be retrieved in the larger systems. Performing three simulations with 128, 256 and 512 processors, we obtain a scaling law, which is approximately $t \propto t(n_{\text{cpu}} = 128)/n_{\text{cpu}}^{0.4}$. Given the few number of simulations, this can of course only give a rough estimate.

The several blocks of the code that compute the solutions of different several physical parts of the model, e.g. the evolution of the phase field or the balance of Li content, follow a similar behavior. We find that for a large number of processors the elastic block scales worst. At the same time, it requires most of the computation time. For the simulations shown in the graphs of FIG. 3.5, the mean contribution of the elastic part was around 82% of the total simulation time. There are several reasons for this effect. The equation describing balance of linear momentum is the only vector-valued problem, which means that the stiffness matrix is larger by a factor of four in a two-dimensional problem. Further, it is nonlinear due to the decomposition of the strain tensor into compressive and tensile parts and, for the same reason, rather bad conditioned.

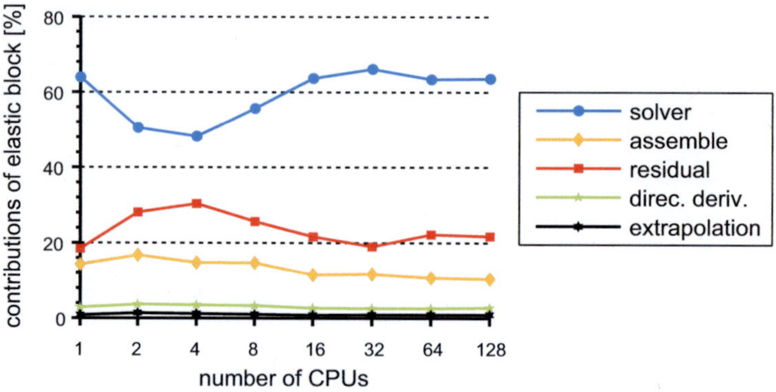

Figure 3.6.: Contributions of different tasks involved in the solution of the equation for balance of linear momentum with increasing number of processors.

The largest share in the computation of the equation describing the balance of linear momentum is taken by the numerical solver itself, as shown in FIG. 3.6. Apart from tweaking certain parameters, there are not many options to reduce this time. However, in comparison to other combinations of preconditioners and solvers, the chosen one performs by far the best. The time required for computing the residual norm is higher than the full assemble step since it is called much more frequently in the iterations to find an optimal step length of the backtracking algorithm.

The contribution of the elastic block becomes more dominant in three-dimensional simulations, as shown in FIG. 3.7. The corresponding geometry in the simulation consists of a half sphere with 15056 elements and 855798 degrees of freedom. As can be observed from the logarithmic scale of the figure, the ratio of the computation time for the elastic block and the total one to run the first 20 time steps is around 96%.

The scaling of the code is excellent for less than 16 and satisfying for less than 128 processors. However, typical three-dimensional simulations exceed the number of elements in this example by up to a factor of 100 and more. To get an impression of the scaling behavior in these cases, we refine the mesh yielding 786 441 elements and approximately 16 million degrees of freedom. To reduce the overall computation time, we run the simulation only for the first five time steps. An almost perfectly linear scaling behavior is obtained from 64 to 96 and 128 processors. Again, this yields only a rule of thumb. However, we can expect that for large simulations such as, for example, described in section 6.2, we should receive a very good scalability for at least up to around 200 processors. Typical simulations are therefore run with 64 to 128 processors in three dimensions and with up to 32 processors in two dimensions.

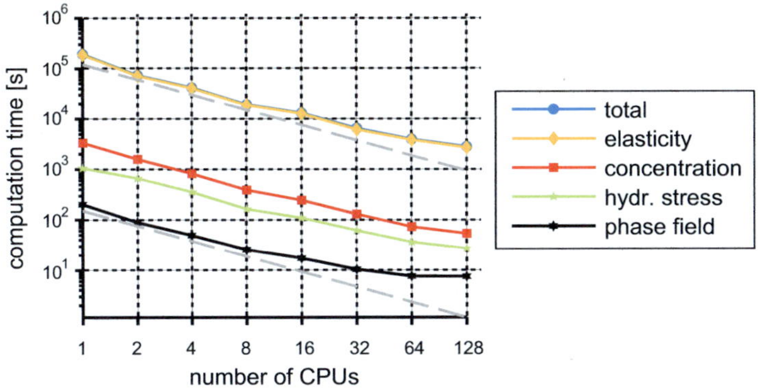

Figure 3.7.: Scaling behavior of the computer code with increasing number of processors. The dashed gray lines illustrate a perfectly linear scaling behavior, i.e. which scales like $t = t_0/n_{\text{cpu}}$.

4 An Assessment of the Phase Field Formulation for Crack Growth

The phase field description for crack growth and fracture is an attractive alternative to numerical methods based on discrete representations of cracks, since the phase field methodology avoids the numerically challenging monitoring of the discontinuities introduced by the crack. In particular, for the simulation of complex crack growth topologies and application to coupled systems, e.g. with thermal or electrical fields, the phase field method has shown promise. However, applying the technique to situations that involve complexities from multiphysical effects, or examining detailed crack patterns that result, for example, from dynamic crack processes, requires validation. This is particularly the case if experiments to visualize the crack trajectory are difficult or impossible to carry out and the researcher therefore has to rely on the results of computations to gain insight into how flaws propagate. The most obvious manner in which validation of the phase field methodology can be achieved is to check whether its results are in agreement with well established results from classical fracture mechanics. The following chapter shall thus present an examination of a typical fracture phase field formulation to check on its conformity with standard results from fracture mechanics. To our knowledge, the closest comparison of this type can be found in [141], but we judge additional results as presented below to be insightful and valuable in regard to the validity of a phase field methodology for fracture.

4.1 Results

Two different setups are chosen to investigate the effectiveness of this particular fracture phase field implementation in solving typical fracture mechanics problems. The first is a pure bending specimen and the second is a load controlled uniaxial stretch specimen, as depicted in FIG. 4.1 a) and c) respectively. Both specimens are studied in two dimensions under plane strain. Further descriptions and analytic fits to numerical solutions for these problems can be found in [161].

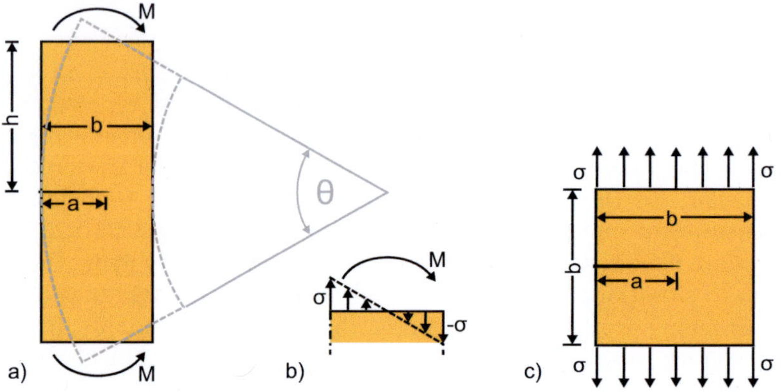

Figure 4.1.: (a) Pure bending specimen with initial crack length a, height $2h$, width b and bending angle θ. (b) Illustration of relation between bending moment M and normal stress σ along upper boundary of the bending specimen. (c) Single edge notch specimen with side length b and normal stress σ.

In both specimens an initial edge crack with length a is introduced. Two different approaches are examined for this step. In one approach, the crack is modeled with free boundaries starting at the edge and reaching into the interior of the mesh, as, for example, utilized in [140]. We will denote this approach as *mesh induced initial crack*. An example of it is shown on the left hand side of FIG. 4.2.

Alternatively, we prescribe the crack through the phase field by defining an initially non-zero history field of the form

$$\mathcal{H}_0 = \alpha\, e^{-(y/\beta)^2} \times \begin{cases} 1, & x < a \\ e^{-((x-a)/\beta)^2}, & x \geq a, \end{cases} \tag{4.1}$$

for a crack that is located at $y = 0$ and extends from the edge at $x = 0$ to $x = a$. The phase field is then allowed to relax in a number of iterations to the given history field, combined with an adaptive re-meshing of the initially uniform mesh. The idea behind this approach is to develop a phase field that resembles the form that it commonly takes on when it describes a crack that has grown due an external load. In those situations, the history field $\mathcal{H}$ at the crack is typically a very localized field with large values at the peak location of the phase field and a steep decrease away from it. The amplitude in equation (4.1) is therefore set to $\alpha = 10^4$ and the width $\beta = l/10$.

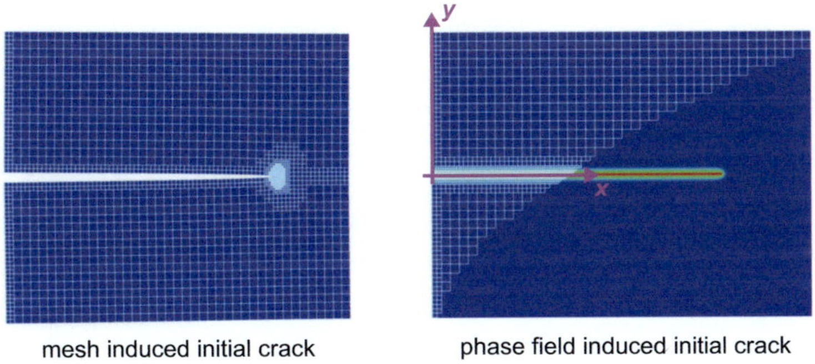

Figure 4.2.: Different approaches for introducing the initial crack. Both figures are depicted before the crack begins to grow. (Left) *Mesh defined initial crack*, modeled with free boundaries of the mesh. (Right) *Phase field induced initial crack*, introduced through the initial history field $\mathcal{H}_0$.

In contrast to the approach in [141], directly setting the initial phase field $d_0(\mathbf{x})$ is not applicable in our formulation. This results from the fact, that the irreversibility condition only enters the history field $\mathcal{H}$. An initially zero history field would therefore always lead to a vanishing order parameter d, independent of its initial condition. The method is also more versatile than using nodal constraints of Dirichlet type for the order parameter since in this case the location of the crack would have to coincide with nodes of the mesh. Further, it can be generalized to model more complex initial crack geometries but a discussion of the details is beyond the scope of this work.

We believe our approach to be a reasonable since the form of the resulting phase field in regions that are induced by $\mathcal{H}_0$ is identical to that obtained when the phase field has grown because of tensile stress. This can be seen in FIG. 4.3 where we plot the value of the order parameter d at a location, where a crack is initially induced through $\mathcal{H}_0$ and at a location through which the crack grows thereafter. The difference is negligible. In addition, the influence of the parameters α and β on the actual shape of the phase field proves to be negligible as long as the amplitude α of $\mathcal{H}_0$ is high enough and its decay, controlled by beta, is faster than the one from the phase field for a grown crack. An alternative realization of the history variable induced precrack can be found in [117]. This approach will be called the case of a *phase field induced initial crack*. It is depicted on the right hand side of FIG. 4.2.

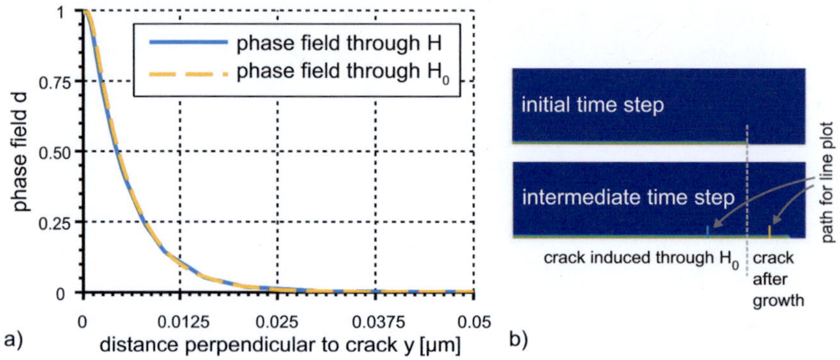

Figure 4.3.: a) Decay of the order parameter d along a line perpendicular to the crack path taken at two locations along the crack, as marked in b). At the first location, the phase field is initially induced through H_0. The second location is somewhere between the initial and final crack tip position, i.e. where the phase field is driven through elastic energy growing the crack.

The specimens are slowly bent or stretched, with the angle or load being proportional to time. The main focus then lies in finding the characteristic critical quantities, such as applied stress or imposed angle, that lead to growth of the crack, and comparing them with analytic results from the literature for quasistatic behavior. The quasistatic growth condition for a crack under mode I loading is such that the relevant stress intensity factor

$$K_I = \sigma\sqrt{\pi a}\, F(a/b) \tag{4.2}$$

must be equal to the plane strain fracture toughness K_{Ic}. The characteristic stress measure σ and shape factor $F(a/b)$ are specific to the respective specimen and will be given later. In terms of energy release rates, this means that if

$$G = \frac{K_I^2}{E'} \tag{4.3}$$

reaches the toughness G_c, the crack starts to extend. For plane strain conditions, the parameter E' is defined as $E' = E/(1 - \nu^2)$. The quantities are thus compared with the relations (4.2) and (4.3), calibrated to the specific specimens investigated.

In most situations investigated, crack growth takes place in an unstable fashion. This means that as soon as the phase field starts to evolve and the crack extends, the local tensile strain energy at the crack tip does not decrease fast enough to arrest crack growth and a new equilibrium cannot be found. Hence, although the applied boundary conditions remain fixed, the crack continues to propagate. Its extension per time step is therefore only controlled by inertia. As a result, the number of alternating iterations between the elastic and phase field computations, as described in section 3.3.4, exceeds a defined threshold. Due to the time adaption algorithm, the time step consequently decreases drastically by orders of magnitude. After a number of time adaption steps, the extension of the crack per time step becomes small enough, so that the prescribed maximum number of iterations for the elastic and phase field computations is sufficient to find the current dynamic equilibrium, and the time step stabilizes again. We take this rapid drop in the time step to be an objective criterion for determining the instant of crack growth initiation and for selecting the respective critical quantities. Since this drop does not occur during stable crack growth, we also exploit it for differentiating between the onset of stable and unstable crack extension.

For crack growth initiation, the length of the crack is defined by the method of introducing it. For the *mesh induced initial crack*, it is given through the length of the free surface; for the *phase field induced initial crack*, it is prescribed by a in the definition of the initial history field, given in equation (4.1). To measure the crack length during crack growth, as in the stable growth regime for the bending test, a different definition of the crack length has to be found. The crack is therefore taken to terminate with its tip at the location where $d = 0.95$. The crack length is measured accordingly and the corresponding critical values of parameters are extracted from the output with this given definition of crack length. We concede that the chosen value of the order parameter to define the crack tip is somewhat arbitrary. Borden et al. [117], for example, use a value of $d = 0.75$ in their work to find the crack tip when identifying its velocity. Furthermore it is not fully consistent with the definition of the crack length for crack growth initiation. However, we emphasize that the ambiguity in defining the crack tip is an intrinsic property of the phase field formulation for fracture and discuss its consequences in section 4.2.

4.1.1 Pure Bending Specimen

The pure bending specimen is selected as a test case since it allows the study of both stable and unstable crack growth. The characteristic stress of equation (4.2) is defined through the bending moment, as illustrated in FIG. 4.1 b), by

$$\sigma = \frac{6M}{b^2} \tag{4.4}$$

and the shape factor is taken from [161] as

$$F(a/b) = \sqrt{\frac{2b}{\pi a} \tan \frac{\pi a}{2b}} \frac{0.923 + 0.199(1 - \sin \frac{\pi a}{2b})^4}{\cos \frac{\pi a}{2b}}, \tag{4.5}$$

giving better than 0.5% accuracy in K_I in terms of equation (4.2) for any a/b. Inserting this function into (4.2) yields a right hand side, which is monotonically increasing with crack length. This means that for a given critical stress intensity factor, the stress necessary to generate crack growth is monotonically decreasing with the crack length. A moment or force controlled bending test thus always leads to a situation of unstable crack growth. In contrast, if the bending is prescribed through the angle, both a stable and an unstable regime can be found.

The total bending angle can be separated into two parts, a bending that assumes a perfect rectangular specimen without a crack and a contribution due to the crack

$$\theta = \theta_{\text{uncracked}} + \theta_{\text{crack}}. \tag{4.6}$$

We prescribe the angle on the upper boundary for which $h/b = 3$. Since the aspect ratio $h/b = 3$ enables use of beam theory, the first part can be deduced to be

$$\theta_{\text{uncracked}} = \frac{\sigma}{E}(1-\nu^2)\frac{4h}{b}, \tag{4.7}$$

For the crack contribution we use the solution from [161], which comes with an accuracy of better than 1% for any a/b. It reads

$$\theta_{\text{crack}} = \frac{\sigma}{E}(1-\nu^2)\,4\,S(a/b), \tag{4.8}$$

with

$$S(a/b) = \left(\frac{a/b}{1-a/b}\right)^2 \sum_{i=0}^{4} q_i (a/b)^i \tag{4.9}$$

and $q_0 = 5.93$, $q_1 = -19.69$, $q_2 = 37.14$, $q_3 = -35.84$ and $q_4 = 13.12$. Combining equations (4.6 – 4.9) and replacing the stress by equations (4.2) and (4.3), we obtain

$$\theta_c = \frac{4}{\pi a}\frac{1}{F(a/b)}\left(\frac{h}{b} + S(a/b)\right)\sqrt{\frac{G_c}{E}(1-\nu^2)} \tag{4.10}$$

as the equilibrium angle at which the energy release rate is equal to the critical one given a crack length a. The dimensions of our specimen are set to $b = 1\,\mu\text{m}$ and $h = 3\,\mu\text{m}$, Young's modulus is $E = 93\,\text{GPa}$, the Poisson ratio $\nu = 0.3$ and the material density $\rho = 4140\,\text{kg/m}^3$. The critical energy release rate is defined by $G_c = 1.2\,\text{J/m}^2$. The corresponding curve is plotted in both FIG. 4.4 and FIG. 4.5. The angle is prescribed through a multi-point constraint. For the upper boundary, the vertical displacement of the center node u_y^0 located at $x = 0$ and $y = h$ is unconstrained and vertical displacements of all other nodes are fixed by $u_y(x) = u_y^0 - \frac{\theta}{2}x$. The edge is therefore free to move vertically while keeping the prescribed angle, so that no net axial force acts on it. The boundary condition on the lower edge is applied correspondingly. The rate of the bending angle was set to $\dot{\theta} = 10^{-4}\,\text{rad/s}$. The loading history can therefore be considered to be quasi-static.

We observe that under rotation controlled boundary conditions, a regime of stable crack growth exists for $a/b \geq 0.67$. In this region, a crack that has grown by a distance Δa, only further extends if the rotation is increased. In contrast, for smaller ratios of a/b, a crack that has once started to grow, does not stop until it reaches the corresponding angle in the stable region at which the energy release rate is again $G = G_c$. This process is depicted in FIG. 4.6. From our experience, unstable crack propagation can only be predicted accurately by a dynamic analysis. Otherwise, an erroneous evolution of the phase field may occur, e.g. by a significantly larger smoothing of d than expected from the length scale parameter. This broadening effect has also been mentioned in [140].

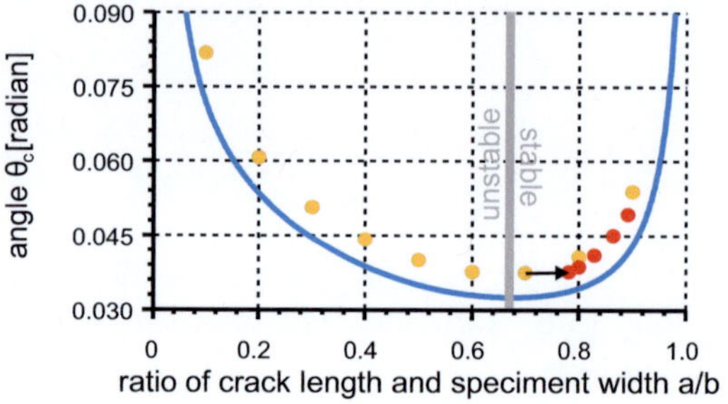

Figure 4.4.: The solid blue line shows the equilibrium angle at which $G = G_c$ depending on the ratio of crack length a to specimen width a/b, as obtained from equation (4.10). Yellow dots mark points of crack initiation obtained from phase field simulations with a *mesh induced initial crack.* The red dots were extracted during stable crack growth from an initially unstable simulation (yellow dot with arrow). The gray line describes the theoretical boundary between stable and unstable growth.

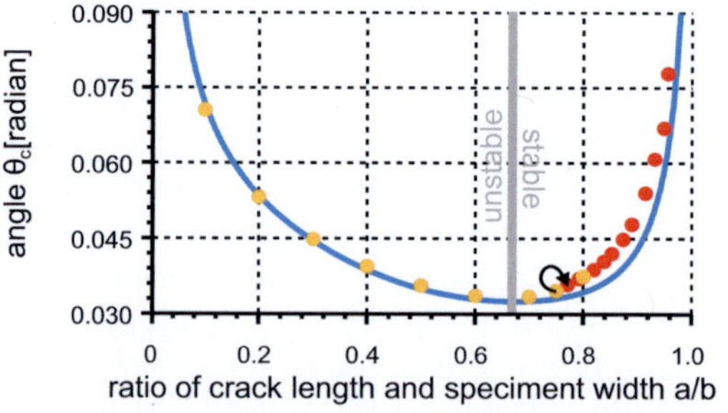

Figure 4.5.: The solid blue line shows the equilibrium angle at which $G = G_c$ depending on the ratio of crack length a to specimen width a/b, as obtained from equation (4.10). Yellow dots mark points of crack initiation obtained from simulations with a *phase field induced initial crack.* The red dots were extracted during stable crack growth from an initially stable simulation (yellow dot with arrow). The gray line describes the theoretical boundary between stable and unstable growth.

As a second quantity to evaluate the predictions on crack growth obtained from the phase field simulations, we compute the energy release rate G for the data points presented in FIG. 4.4 and 4.5. We combine equations (4.2) and (4.3) yielding

$$G = \sigma^2 \, \pi a \, F^2(a/b) \tag{4.11}$$

and insert the measured crack length and applied stress σ, as given by equation (4.4). The results are normalized by the input parameter G_c and plotted in FIG. 4.7.

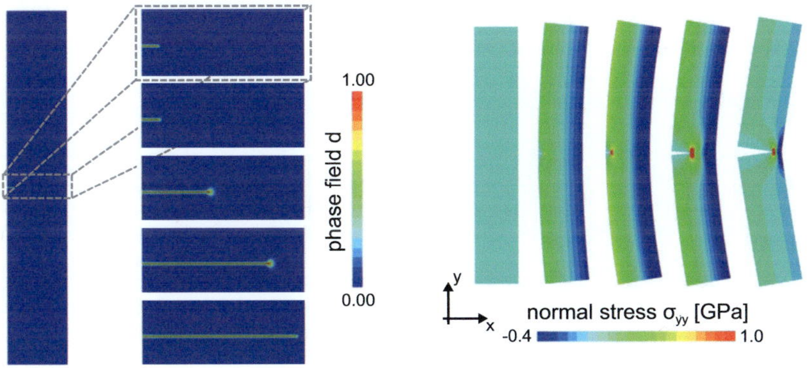

Figure 4.6.: Illustration of results for the bending specimen. (Right) Normal stress in y-direction is plotted for a number of bending angles. The displacements are scaled by a factor of 5 to show the deformation of the specimen exaggerated. (Left) During bending an initially small crack propagates in an unstable fashion and stops at a location close to the opposite free edge. An order parameter of $d = 1$ marks cracked regions, whereas $d = 0$ corresponds to intact regions.

4.1.2 Single Edge Notch Specimen

To strengthen the findings on the predictions of the phase field method for crack growth obtained from the observations on the pure bending specimen, a second series of simulations is run for a single edge notch specimen with force controlled boundary conditions. The corresponding geometry, shown in FIG. 4.1 c), represents the most simple setup to study crack growth. Due to its unstable behavior at crack initiation, it allows determination of the critical stress very precisely and is therefore suitable for examining the influence of the length scale parameter l on the numerical results. Other works, investigating this kind of specimen, can be found in, for example, [140, 141].

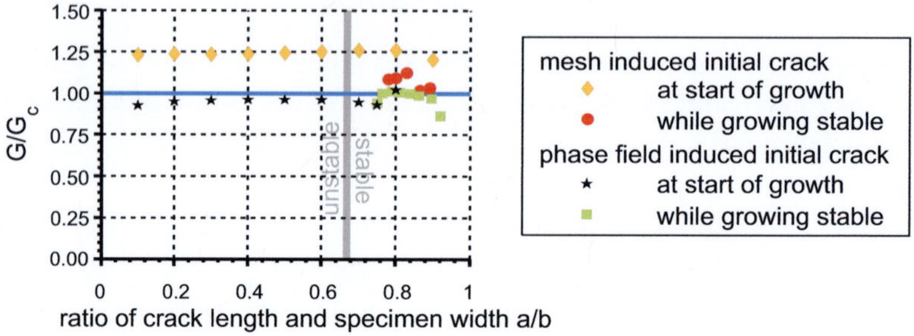

Figure 4.7.: Computed energy release rate G normalized by the critical energy release rate G_c. Data points correspond to those shown in FIG. 4.4 and FIG. 4.5. Again, the gray line describes the theoretical boundary between stable and unstable growth. *Mesh induced initial crack*: The start of crack growth (yellow diamonds) is always unstable. The red dots indicate the values of G for cases where the growth of a *mesh induced crack* has changed to a stable regime. *Phase field induced initial crack*: The start of crack growth (black stars) is unstable or stable in agreement with the theoretical boundary. The values of G computed during stable crack growth are denoted by the green squares.

The characteristic stress of equation (4.2) is equal to the normal stress in the y-direction $\sigma = \sigma_{yy}$ at the upper boundary and the shape factor is again taken from [161]

$$F(a/b) = \sqrt{\frac{2b}{\pi a}\tan\frac{\pi a}{2b}}\,\frac{0.752 + 2.02(a/b) + 0.37\left(1 - \sin\frac{\pi a}{2b}\right)^3}{\cos\frac{\pi a}{2b}}, \quad (4.12)$$

with better than 0.5% accuracy in $F(a/b)$ for any a/b. The width of the specimen is $b = 1\,\mu\text{m}$, Young's modulus is set to $E = 93\,\text{GPa}$, Poisson ration to $\nu = 0.3$, the material density to $\rho = 4140\,\text{kg/m}^3$ and the critical energy release rate to $G_c = 1.2\,\text{J/m}^2$.

The load on the specimen is increased slowly with time until the crack starts to grow. We again take the drop in the time step resulting from the time adaption scheme to determine the moment of crack growth ini-

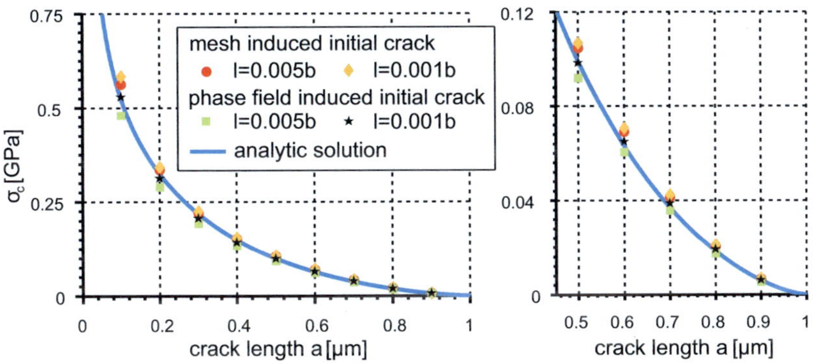

Figure 4.8.: Applied external stress for varying crack length a at instance when $G = G_c$. Colored markers are stress values taken from simulations at the moment when the crack starts to grow. The diagram on the right is a magnification of the points with larger initial crack length.

tiation in the phase field simulation. The corresponding critical stress is presented in FIG. 4.8 for both *mesh* and *phase field induced initial cracks*, for different initial crack lengths a and two values of the length scale parameters l. The results are compared with the theoretical stress at which $G = G_c$. For the latter, equations (4.2) and (4.3) are combined to give

$$\sigma_c = \sqrt{\frac{G_c E}{1-\nu^2}} \frac{1}{\sqrt{\pi a}} \frac{1}{F(a/b)}. \tag{4.13}$$

We impose the critical stress as boundary condition in a second, purely elastic simulation without phase field to determine the energy release rate using the J-integral method. For the evaluation of J, we use the commercial tool COMSOL Multiphysics®. For the presented setting J is a path-independent line integral. Given a counterclockwise path S around the tip of a crack, it reads

$$J = \int_S \left(w\,\mathrm{d}y - t_i \frac{\partial u_i}{\partial x} \mathrm{d}s \right), \tag{4.14}$$

with the length increment $\mathrm{d}s$ along the contour Γ, the strain energy density $w = \int_0^{\varepsilon_{ij}} \sigma_{ij} \mathrm{d}\epsilon'_{ij}$ and the traction vector $t_i = \sigma_{ij} n_j$ in normal direction $\mathbf{n} = n_i \mathbf{e}_i$ to the path [132]. For the given setting of a linear elastic material under quasi-static conditions, the magnitude of the J-integral is equal to the energy release rate, $J = G$. The so computed energy release rates are plotted in FIG. 4.9.

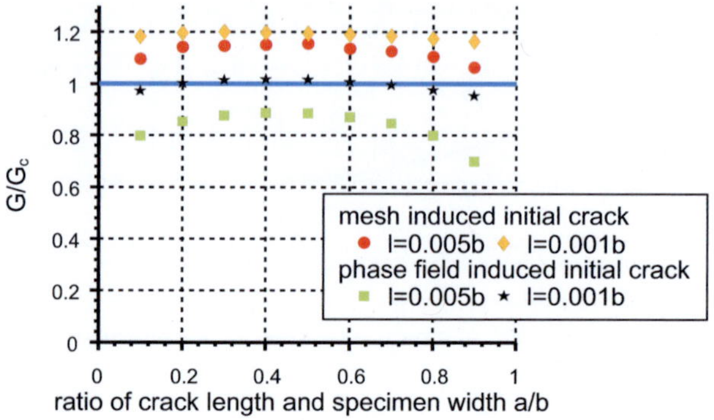

Figure 4.9.: Normalized energy release rates G/G_c at crack growth initiation computed via the J-integral method against initial crack length a for two values of the length scale parameter l and both methods of introducing the initial crack.

In order to explore the influence of the length scale parameter l on the predictions of the phase field method more closely, simulations with l ranging from $0.005\,b$ to $0.0005\,b$ are run for an initial crack length of $a = b/2$, including both methods of introducing the initial crack. The resulting energy release rates are shown in FIG. 4.10.

4.2 Discussion

4.2.1 Pure Bending Specimen

Comparing the critical angle extracted from simulations with the analytic solution (cf. FIG. 4.4) in the unstable regime, we observe a non-negligible discrepancy between 13.3% and 14.6% for computations with a *mesh induced initial crack*. As will be shown for the single edge notch specimen, this discrepancy does not result from the mesh or the staggered solution scheme. We attribute it to an initial build-up phase required to create a region of $d = 1$ at the crack tip, that significantly postpones the crack initiation in the sense of an evolving phase field.

In contrast, the numerical results for the *phase field induced initial crack* are in very good agreement with the analytic ones for the case of unstable crack growth, see FIG. 4.5. The error in θ_c is always below 2.1% in the unstable regime.

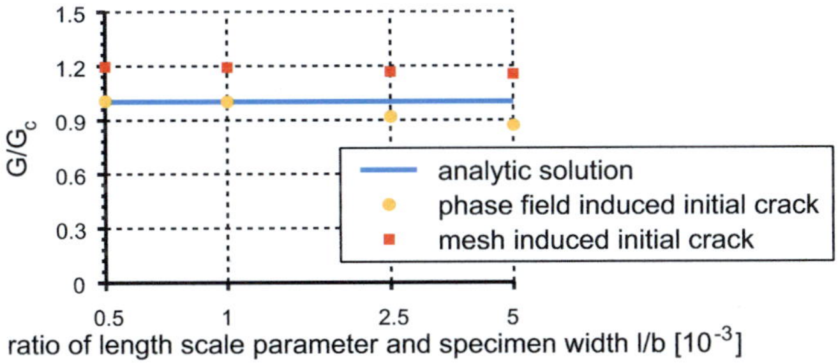

Figure 4.10.: Normalized energy release rates G/G_c computed via J-integral method against ratio of length scale parameter l and specimen width b for both methods of introducing the initial crack.

In the stable regime, the results from the two methods of introducing the crack are closer to each other, when the data is extracted during crack growth. This agrees with our expectations, since in both cases the growth of the crack is described through the same extension of the non-zero phase field region. However, there is a slight offset between numerical and analytical data, even for the *phase field induced initial crack.* One reason for this difference might lie in the non-unique way that the length of a phase field crack may be measured during crack growth, resulting from the continuous increase of the order parameter at the crack tip from zero to unity. To make this point clearer, some numbers shall be given here. For all bending specimen calculations, the crack tip was defined to be located at $d = 0.95$. However, if we measure the crack tip instead at $d = 0.5$ ($d = 0.1$), the crack length would be longer by $\Delta a \approx 0.005\,b$ ($\Delta a \approx 0.018\,b$). To estimate, how large the effect of the criterion on the error in terms of the bending rotation is, we compute the Taylor expansion of equation (4.10) to first order in a yielding

$$\Delta\theta_c = \left|\frac{4}{\pi a}\frac{1}{F(a/b)}\left[-\left(\frac{1}{a}+\frac{1}{b}\frac{F'(a/b)}{F(a/b)}\right)\left(\frac{h}{b}+S(a/b)\right)+\frac{1}{b}S'(a/b)\right]\right| \times\sqrt{\frac{G_c}{E}(1-\nu^2)}\,\Delta a. \tag{4.15}$$

For a crack length of $a = 0.914\,b$ and an error of the crack length of $\Delta a \approx 0.005\,b$ ($\Delta a \approx 0.018\,b$), taken from a simulation with a phase field induced precrack, this gives a relative error in the angle of around $\Delta\theta_c/\theta_c \approx 5.4\%$ ($\Delta\theta_c/\theta_c \approx 19.5\%$). The actual relative error from a phase field computation with a crack of this length is $|\theta_c^{\text{num}} - -\theta_c^{\text{ana}}|/\theta_c^{\text{ana}} \approx 18\%$. The uncertainty in measuring a is therefore believed to contribute a significant fraction of the discrepancy between the results from the phase field calculation and that from the fracture mechanics analysis of the problem based on equation (4.2), (4.4) and (4.5). This issue is also further discussed in section 4.2.2 for the stretch specimen. In the unstable regime, we observe good agreement for the *phase field induced crack*. The energy release rate is only slightly underestimated, by less than 10%. In contrast, for the *mesh induced initial crack*, G deviates from G_c by around 25%. In the stable region, both methods of introducing the crack give results that lie very close to each other.

A second contribution to the discrepancy may come from finite values of the order parameter in regions far away from the crack. It is notable that the order parameter is perceptibly above zero in the ligament ahead of the crack tip. For example, in the example mentioned above with $a = 0.914\,b$, the order parameter is $d = 0.01$, if extracted at the compressive fiber on the right hand side of the specimen in the extension of the crack. Treating d as a damage parameter in the sense that we can define an effective elastic modulus given by $E^* = (1-d)^2 E$ [162], this would mean a decrease of around 2% for the stiffness of the material. Given that this location is more compliant, the applied moment required to bend the specimen by a certain angle is decreased. This means, that the actual stress acting at the crack tip, is lower than expected for a given bending angle. For the simulations with a *mesh induced initial crack*, the value of d is larger by around 5% at the free surface beyond the ligament leading to a decrease in stiffness of the same order. It is difficult to estimate if this small difference can explain the slightly larger offset for the *mesh induced initial crack*. An additional influence could come from the initially unstable crack growth even in the stable regime, as described below. The growth velocity affects the shape of the phase field, as for example mentioned in [117], and could therefore also have an impact on the results.

Interestingly, although the critical bending angle is generally overestimated for these points, the energy release rate is quite accurate. We ascribe this to a compensating effect due to the underestimation of the bending moment, as described above. The non-zero values of d at presumably intact regions, leading to this underestimation, are strongly dependent on the history of the calculation due to the irreversibility condition in $\mathcal{H}$. For the *mesh induced initial crack*, where crack initiation generally starts later than it should, regions away from the crack may experience higher stress than for the *phase field induced initial crack*. Hence, in this case, the order parameter should be slightly larger in intact regions. To verify this hypothesis, we pick one simulation for each method of describing the initial crack, find the times where the crack has grown to the same extent and study the magnitude of d in regions away from the crack. We find, that for the *mesh induced initial crack* the order parameter is larger by around 3-5% in intact regions. This may lead to the larger offset of the data points associated with stable growth in FIG. 4.4 compared to FIG. 4.5. The argument might also explain the discrepancy in computing G, if we assume that the overestimation of the crack length a, as defined through the order parameter d, has a more dominant effect than the underestimation of σ.

Two mechanisms leading to deviations of the numerical results are identified, that apply to the situation of a *phase field induced initial crack* or when the crack is already growing from a *mesh induced initial crack*. A global one stemming from a non-zero phase field at locations that should actually be perfectly intact and a local one from the inaccuracy of measuring the crack length. Both issues, in particular the latter one, can be eased by choosing a smaller length scale parameter l. As described above, for the used value of $l = 0.005\,b$, the outcome of the simulations is in good agreement with solutions from classical fracture mechanics.

In contrast, crack initiation for a *mesh induced initial crack* occurs always with a significant delay. Both the energy release rate and the critical bending angle are therefore overestimated. An overestimation of the critical failure load was also reported by Bourdin [163], who presents a backtracking algorithm in order to overcome this deficiency. We believe this to be a result of an initial build-up phase, which the phase field has to undergo before true crack growth can initiate. As a matter of course, this build-up is not required if the initial crack is already described through the phase field. A critical consequence of the initial overestimation of G_c is a predicted crack growth that is always initially unstable. This can also be observed in FIG. 4.4. The yellow data point at the tail of the arrow, used as the starting point for the stable results depicted by red, gives an initial spurt of crack growth that directly jumps to the first stable result (in red) without significant change of the bending angle (horizontal arrow). This behavior is not found with the *phase field induced initial crack.* In this case, the simulation predicts entirely stable crack growth when the initial crack length is in the stable regime. For this reason, the yellow and red data points in FIG. 4.5 lie on the same curve. This finding undermines the claimed capability of the fracture phase field method to find and accurately describe crack initiation at arbitrary locations in a specimen.

4.2.2 Single Edge Notch Specimen

The most accurate results in terms of the predicted critical stress, shown in FIG. 4.8, are obtained in this case with the *phase field induced crack* and $l = 0.001\,b$. The relative error $\Delta\sigma_c = (\sigma_c^{\text{phase field}} - \sigma_c^{\text{analytic}})/\sigma_c^{\text{analytic}}$ of the critical stress $\sigma_c^{\text{phase field}}$ resulting from the phase field simulations with respect to the calculated one, $\sigma_c^{\text{analytic}}$, from equation (4.13) ranges between 0.4% and 5.2%. Computations with $l = 0.005\,b$ generate predictions with a similar accuracy for both the *phase field* and *mesh induced crack* of up to 11.8% and 11.3% respectively. However, the least accurate results of $\Delta\sigma_c \approx 14.5\%$ are obtained from the *mesh induced crack* with the smaller value of the length scale parameter $l = 0.001\,b$. This behavior is unexpected since it does not allow the improvement of predictions from phase field simulations by choice of numerical parameters that result in a better shape of the phase field more closely resembling the discrete nature of a crack. This is further discussed in terms of energy release rates G below.

As can be seen in FIG. 4.10, the predictions of the phase field simulations with respect to the obtained value of G are quite satisfactory for the case of a *phase field induced initial crack* and a small length scale parameter l. This was already reported in [141], see in particular FIG. 6 of their work. For example, the computed energy release rate G for $l = 0.001\,b$ is only 1.6% below the critical one G_c. For the corresponding simulations with different initial crack lengths a, the relative error $\Delta G = (G - G_c)/G_c$ ranges between 0.2% and 4.5%. The energy release rate is slightly underestimated for very small and large cracks, as shown in FIG. 4.9.

The tendency is the same for $l = 0.005\,b$, but the results become less satisfactory. The energy release rate G lies around 11% to 30% below G_c. This underestimation is again attributed to the overestimation of the crack length due to the diffusive representation of the crack in the phase field method. For example, in the simulations with $l = 0.005\,b$, the phase field drops to $d = 0.5$ ($d = 0.1$) at a distance of $\Delta x = 0.011\,b$ ($\Delta x = 0.027\,b$) from the crack tip when crack growth is just about to start. By varying the crack length in J-integral computations, we find that a difference in the crack length of $\Delta a \approx 0.02\,b$ can lead to an approximate change in G of 17%. This is in the range of the relative error ΔG.

The effect is less pronounced for cracks close to half the width of the specimen. For small cracks, the slope of the critical stress σ_c against the crack length becomes larger (see FIG. 4.8) so that small differences in the crack length have a stronger impact on the resulting σ_c. In the case of large cracks, the relative inaccuracy of the width $\Delta(b - a)$ of the ligament connecting the upper and lower part is increasing, which makes the computation of the critical stress less precise.

The simulations with a *mesh induced initial crack* show a similar, though slightly attenuated behavior. However, as before, there is a noticeable offset of 16% to around 20% between the computed energy release rates and the critical one for $l = 0.001\,b$. Moreover, the discrepancy is even intensified for smaller values of the length scale parameter l, as observable in FIG. 4.10, so that convergence towards the actual value of G_c is not observed. As for the pure bending specimen, we attribute the overestimation of G_c to an initial build-up of a non-zero phase field region. This presents an intrinsic drawback of the formulation since it postpones the actual crack initiation. To exclude any influence of the mesh, e.g. by not sufficiently resolving the singular stress field at the crack tip, we repeatedly reduce the element size around it. Even for an element size of $h \approx 6 * 10^{-6}\,b$ around the crack tip an improvement in terms of an earlier crack growth initiation is not recognizable.

To ensure that the error is not an outcome of the staggered implementation scheme, we extract data from a similar simulation with a monolithic solution scheme, found in FIG. 12 of [109]. In contrast to our own simulations, the results of the corresponding work are generated with displacement controlled boundary conditions. We insert the critical displacements u_c taken from the diagrams as boundary conditions and compute the energy release rate via the J-integral. The results overestimate G_c by a factor of 19% to 24%, depending on the critical displacements (input parameter is $G_c = 2.7 * 10^3\,\mathrm{J/m^2}$, extracted energy release rates are $3.22 * 10^3\,\mathrm{J/m^2}$ and $3.35 * 10^3\,\mathrm{J/m^2}$). This is even slightly higher than our results.

4.3 Conclusion

In this chapter, a study on the accuracy of predictions from a fracture phase field formulation on crack growth initiation is undertaken and compared with relations from fracture mechanics. We investigate the influence of the numerical length scale parameter l, determining the width of smoothing of the represented crack, and two methods of introducing an initial crack. Two setups are examined, a pure bending specimen and a single edge notch specimen. Due to the requirement of a fine mesh for small values of l, the model is implemented in a parallelized finite element computer code supporting both mesh and time adaption.

We find that results of the phase field simulations are in very good agreement if l is small and the initial crack is induced as a heterogeneity in the phase field through an initially non-zero history field (*phase field induced initial crack*). Predictions become less satisfying for large values of the length scale parameter l. In situations, where the initial crack is prescribed through free surfaces of the mesh (*mesh induced initial crack*), predictions from the phase field simulations deviate significantly from analytic relations found in the literature. This is neither a consequence of an overly coarse mesh, nor of the staggered implementation scheme, and can also not be resolved using a smaller value of l.

Due to the advantages of the phase field approach for simulating complex crack topologies and in coupled applications with multiphysical phenomena, the method is worth considering as an alternative or in addition to established discrete fracture techniques. Accurate results should be obtained if initial cracks are introduced in the manner described above, i.e. with a phase field induced initial crack with a small value of l. However, small values of the length scale parameters l pose computational demands that must not be underestimated and the step of reducing the numerical effort by choosing larger values of l must be taken with care. Considering the results from the *mesh induced crack*, the capability of the presented fracture phase field formulation to find and accurately describe crack initiation at arbitrary locations in a specimen is disproved. To overcome this difficulty, further development of the method is required.

5 A Study on the Role of the Length Scale Parameter as a Material Parameter

In the absence of cracks, classical fracture mechanics states that crack growth cannot occur since the critical load required to initiate crack growth becomes infinite with the crack length a going to zero. However, in simulations of specimens under an externally applied load, which utilize a phase field method for fracture, it is often observed that the order parameter d increases at the locations with highest mechanical stress even if it is initially zero throughout the entire specimen and no crack has been introduced in the material. It is therefore reasonable to assume that a specific type of initial crack distribution can be attributed to the phase field.

The parameters entering the evolution equation of the phase field 2.29 form the dimensionless group $\frac{\mathcal{H}}{G_c/l}$, which represents the driving force for crack growth in the phase field approach. Among the parameters influencing the evolution of the phase field, the length scale parameter l is the only one which is not directly linked to a physical parameter, such as, for example, the critical energy release rate G_c, and which possesses the dimension of a length. Hence, the imagined distribution of initial cracks is assumed to be connected to the value of the length scale parameter l. As a consequence, the length scale parameter should not only be considered as a numerical parameter, but instead might act as a material parameter. Investigations on this hypothesis, e.g. about the influence of the length scale parameter on the critical stress to induce fracture, are presented, for example, in [112, 117, 164, 165] by means of both homogeneous and non-homogeneous analytic solutions in one dimension and numerical tests in two dimensions. Here, we examine a set of simple two-dimensional test specimens and derive relations between the length scale parameter l and an imagined initial crack length a. Using these relations, we explore whether it is a reasonable and practical approach to treat l as a material parameter in order to represent an initial crack length or strength of the material. The following study shall thereby also serve as a clarification on whether this approach is applicable to the fully coupled model, describing crack growth and Li diffusion.

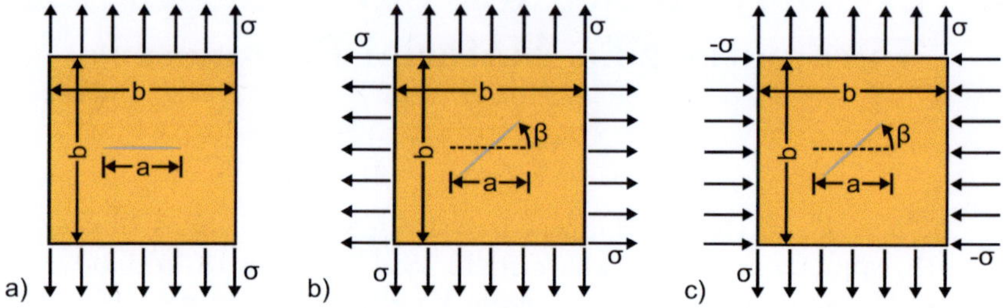

Figure 5.1.: Geometry and boundary conditions for a) uniaxial stretch b) biaxial stretch and c) pure shear specimen. The crack in light gray is only shown for better understanding of the considerations on the orientation of the imagined initial crack. It is not introduced in the actual simulation.

5.1 Results

Either uniaxial, biaxial or pure shear load is applied to the rectangular test specimens, as shown in FIG. 5.1. The edge length of the specimens is $b = 1\,\mu$m and material parameters are the same as for the pure bending and single edge notch specimen, presented in chapter 4. In order to save computation time, only the upper half of the geometry is simulated. The external load is increased until the order parameter d raises to unity. This is accompanied by a drastic drop of the time step, as described on page 69 of chapter 4. This drop is used to determine the instant of failure. The critical external stress, extracted at this point in time, is plotted in FIG. 5.2.

To analyze the influence of the length scale parameter l, we imagine the following scenario. The test specimens shall be covered by cracks of length a and orientation angle β with respect to the x-axis. Increasing the applied stress to a certain critical value, the condition is reached where the cracks grow. Speaking in terms of the order parameter, this corresponds to d rising to unity. We thus relate the critical stress obtained from the simulations to an effective length of an imagined crack.

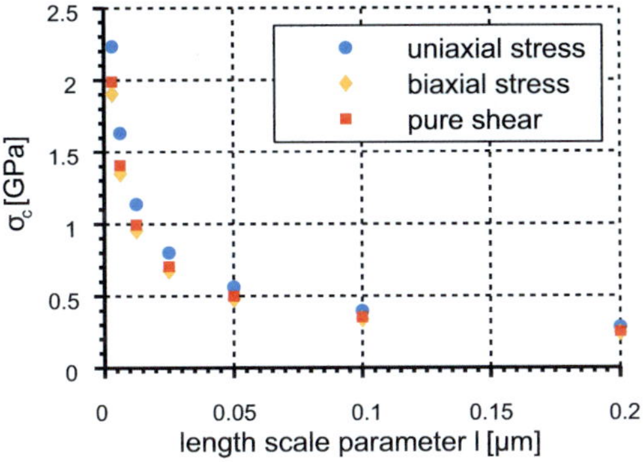

Figure 5.2.: The critical stress σ_c at which the order parameter becomes $d = 1$ in specimens under uniaxial, biaxial and pure shear load, plotted for different values of the length scale parameter.

For this, we assume an orientation angle β of the crack such that the critical stress required to initiate crack growth is minimized. Hence, β is found by maximizing the stress intensity factor. For the uniaxial stress specimen, the mode I stress intensity factor is largest if the crack is aligned perpendicular to the applied load. It is given by equation (4.2). Considering the large critical stress values obtained from the computations, it is reasonable to assume that the effective crack length is very small, i.e. $a \ll b$. In this case, the shape factor $F(a/b)$ converges to a constant, i.e. $\lim_{a/b\to 0} F(a/b) = f$. For a through crack in the interior (an edge crack), the constant becomes $f = 1$ ($f = 1.122$). Within this approximation, equation (4.2) reads

$$K_I = \sigma f \sqrt{\pi a}. \tag{5.1}$$

We isolate the crack length a and replace the stress intensity factor with the energy release rate using equation (4.3). For the instant of crack growth initiation, i.e. when $G = G_c$ and $\sigma = \sigma_c$, we then receive the effective crack length of the imagined crack by

$$a = \frac{1}{\pi} \frac{1}{1-\nu^2} \frac{G_c E}{\sigma_c^2 f^2}. \tag{5.2}$$

By rotation of the coordinate system and application of the principle of superposition [132], it can be shown that equation (5.1) also applies to the biaxial stretch and pure shear specimen, if the crack is expected to lie in the interior, i.e. when $f = 1$. In the case of a biaxial load with equal load from all sides, the stress intensity factor is independent of the orientation angle β due to the purely hydrostatic stress field. In the pure shear test, the most severe angle for mode I loading is again $\beta = 0°$. Considering mode II loading, the stress intensity factor is highest at $\beta = 45°$ with $K_{II} = \sigma\sqrt{\pi a}$ and $K_I = 0$. Since the maximum stress intensity factors are identical for all the test specimens, we use equation (5.2) to determine the effective crack length corresponding to the respective critical stress. The so computed crack lengths are shown in FIG. 5.3.

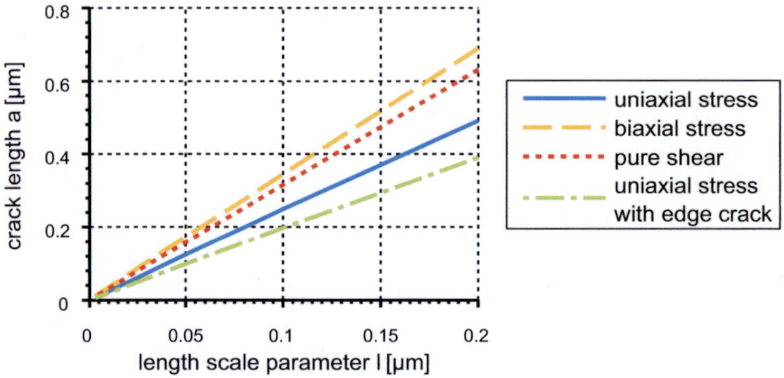

Figure 5.3.: The effective crack length a in specimens under uniaxial, biaxial and pure shear load, plotted for different values of the length scale parameter l.

To confirm our findings in the relation between l and a, we further perform a similar study examining the pure bending specimen described in section 4.1.1. Commencing with an entirely intact specimen, i.e. with $d = 0$ everywhere, we monotonically increase the imposed angle on the upper boundary and extract the critical angle at which the specimen fails, i.e. when the phase field raises to unity at some location in the specimen. We repeat this procedure for several values of l and compute the effective crack length matching the critical angle by numerically solving equation (4.10) for the crack length a. Both critical angle and crack length are shown in FIG. 5.4.

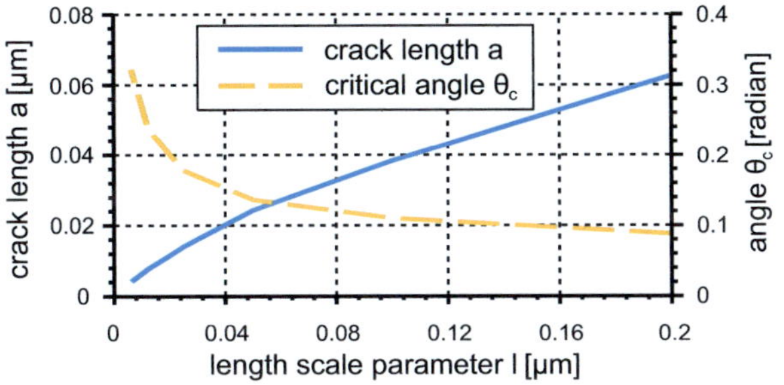

Figure 5.4.: The critical angle θ_c and the effective crack length a, plotted for different values of the length scale parameter l.

5.2 Discussion

The graphs describing the effective crack length a, shown in FIG. 5.3, follow a linear relation with respect to the length scale parameter l. Fitting the plotted data points with a polynomial of degree one obtains the following relationship between l and a

$$\begin{aligned} \text{uniaxial stress with through crack} \quad & a \approx 2 * 10^{-4}\, b + 2.46\, l, \\ \text{uniaxial stress with edge crack} \quad & a \approx 4 * 10^{-8}\, b + 1.95\, l, \\ \text{biaxial stress with through crack} \quad & a \approx 4 * 10^{-8}\, b + 3.45\, l, \\ \text{uniaxial pure shear with through crack} \quad & a \approx 2 * 10^{-4}\, b + 3.16\, l. \end{aligned}$$

Considering the negligible size of the offsets obtained in the fits, we conclude that as l approaches zero, also a vanishes. This is in agreement with the derivation of the phase field method for fracture. For decreasing values of the length scale parameter l, the spatially smoothed representation of the crack converges to the discrete shape. The approximation of the fracture energy by equation (2.23) becomes more accurate and results, obtained using the phase field approach, should resemble more closely predictions from standard fracture mechanics. In this case, a non-zero initial crack length does not exist.

The relations presented above give a rough estimate of the intrinsic, effective crack size obtained when using a certain length scale parameter l. At the same time, they provide a measure for the size of an initial crack that can still be resolved in a fracture phase field simulation with a given length scale parameter. Only if the initial crack, induced, for example, by the methods presented in chapter 4, is noticeably larger than the effective crack length related to the given value of the length scale parameter, the role of l as a material parameter can be neglected. However, we observe that the effective crack length, corresponding to a certain value of the length scale parameter l, depends on the specific type of load or, in general, the boundary conditions. To understand this point better, we examine the behavior of the phase field at failure more closely.

The evolution of the order parameter d follows the same trend for both the uniaxial and biaxial stretch specimen. As the external load is increased, the order parameter first grows homogeneously to a value between $d \approx 0.23$ and $d \approx 0.33$ and then abruptly raises to $d = 1$ at specific locations in the specimen. As can be seen from FIG. 5.5, large values of the order parameter d are localized at certain positions in the specimen even if geometry and boundary conditions should generate a uniform stress field. In the uniaxial and biaxial stretch test, the regions of the highest value of the order parameter are found at the lower and upper edges and corners respectively. We assume that numerical errors slightly increase d at these locations. By a self-enforcing effect, the phase field then grows faster at the respective locations. Remarkably, this behavior is significantly less pronounced in the case of the pure shear specimen, where the order parameter raises above $d \approx 0.82$ everywhere.

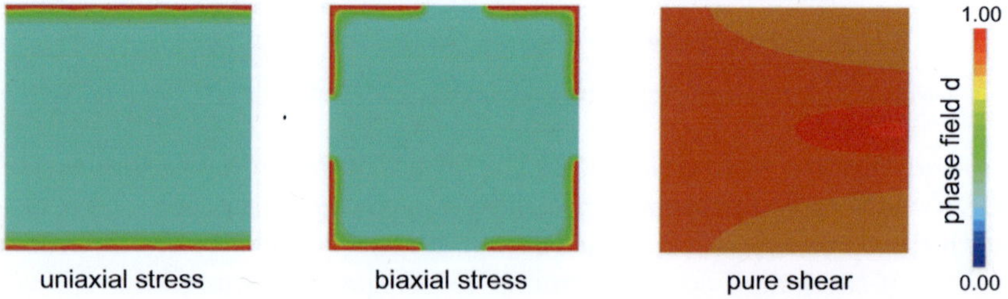

Figure 5.5.: Contour plots of the order parameter at failure in specimens under uniaxial, biaxial and pure shear load.

In an arbitrarily shaped specimen, the stress condition is typically neither known a-priori nor is it as simple as in these specimens. Hence, it is generally not possible to determine which of the presented specimens resembles closest the stress situation in the specimen in question, also because details in the geometry can influence the local stress field in a significant way. Hence, a relation between a and l, or equivalently a l-dependent material strength, cannot be predicted in advance for an arbitrarily shaped specimen. This makes it questionable, if the length scale parameter l can be used in a practical way to define an initial crack length or material strength. Furthermore, the behavior of the phase field before critical conditions are reached does not describe brittle fracture in the actual sense. The homogeneous increase of the order parameter d, preceding the failure of the specimen, leads to a nonlinear increase of the material compliance and is more reminiscent of damage theory, describing a distribution of small cracks in a continuous way without resolving each crack [162, 166].

The linear relation between the crack length a and the length scale parameter l, found in the stretch and shear specimens, is not reproduced in the bending specimen. This can be seen in FIG. 5.4, which shows the critical angle for failure θ of the specimen and the corresponding effective crack length a. Fitting a polynomial of degree one to an increasingly smaller selection of the corresponding data points, we find the following relations

$$\begin{aligned}
&\text{all data points} && a \approx 0.0058\,b + 0.30\,l,\\
&\text{leaving away } l > 0.1 && a \approx 0.0038\,b + 0.36\,l,\\
&\text{leaving away } l > 0.05 && a \approx 0.0019\,b + 0.46\,l,\\
&\text{leaving away } l > 0.025 && a \approx 0.0011\,b + 0.52\,l,\\
&\text{leaving away } l > 0.0125 && a \approx 0.0005\,b + 0.58\,l.
\end{aligned}$$

The proportionality constant between a and l grows and the offset tends to zero, as larger values of l are excluded from the fitting procedure. While the trend of the offset points in the expected direction, the slope of the functions is significantly different in comparison to the ones obtained from the stretch and shear specimens. Whereas the typical initial crack size was found to be between $a \approx 2\,l$ and $a \approx 3\,l$ in the stretch and shear specimens, the results from the bending specimen suggest that it is closer to $a \approx l/2$. To understand this incongruity, we take a closer look at the evolution of the phase field during bending of the specimen. For this reason, we compare the phase field at an instant just before the critical angle θ_c is reached and directly at θ_c, when a crack is about to develop. The respective contour plots are shown in FIG. 5.6. The length scale parameter used in the corresponding simulations is $l = 0.00625\,b$.

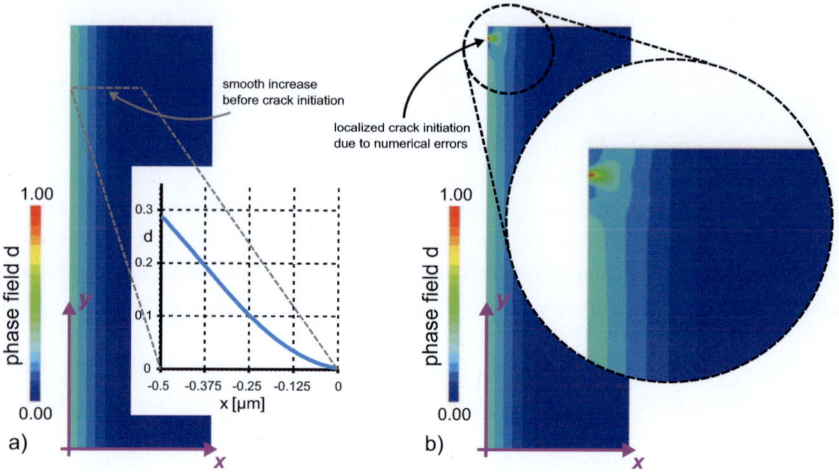

Figure 5.6.: Evolution of the phase field with increasing bending angle drawn a) shortly before a crack starts to grow and b) at crack initiation. The diagram on the left shows the value of the phase field along a horizontal line from the outer edge to the center of the specimen.

We observe that before a crack nucleates and begins to grow, the order parameter already reaches $d \approx 0.3$ at the tensile fiber on the left hand side of the specimen. This non-zero phase field relaxes the stress in y-direction along the edge and results in a nonlinear material response. The nonlinear behavior of the material can also be observed in the normal stress in y-direction at the upper boundary of the specimen, as plotted in FIG. 5.7. In a purely elastic specimen, the prescribed angle results in a torsional moment around the center of the upper boundary. Hence, the normal stress σ_{yy} should be zero in the center of the upper edge and follow a linear relation with respect to the x-coordinate. This behavior is found for the lowest bending angle. However, as the bending angle is increased, the phase field rises at the outermost tensile fiber, resulting in a more compliant material response in this part of the specimen. As a consequence, the location of zero normal stress slightly moves towards the compressive fiber on the right hand side of the specimen and the graph of the normal stress in y-direction no longer describes a straight line.

From these observations, we conclude that the specimen behaves significantly different than expected for a purely elastic material. However, an elastic material response was assumed in the derivation of the analytic relations between effective crack length a and critical angle θ_c. The discrepancy in the relationship between l and a, comparing the stretch and shear specimens to the bending specimen, is thus caused by the damaging character of the phase field in advance to the actual crack growth initiation.

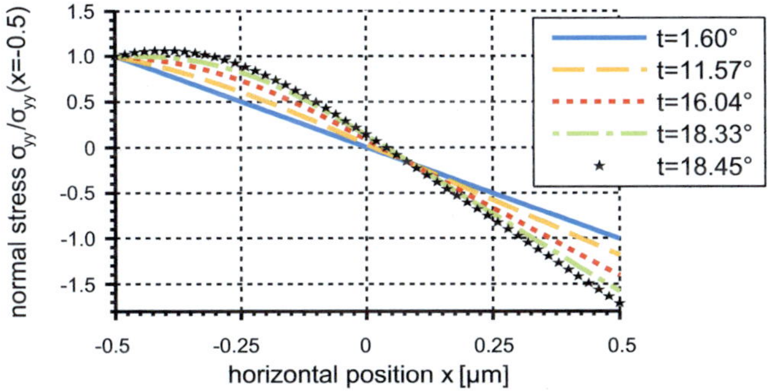

Figure 5.7.: Normalized normal stress in y-direction, plotted in x-direction along the upper boundary of the specimen for different values of the bending angle. See FIG. 5.6 for the direction of coordinate axes. The dot-dashed, green line corresponds to the situation in FIG. 5.6 a), the dark gray stars are taken from the situation illustrated in FIG. 5.6 b).

In order to find a better estimate of the dependence of the effective crack length a on the length scale parameter l in the bending specimen, we assume that in a small area along the outermost tensile fiber, the stress state is roughly comparable to a pure stretch specimen. We therefore extract the normal stress in y-direction, that is generated at the upper left corner at the time when the critical angle is reached, and insert it into equation (5.2). Depending on the selection of data points, we obtain the following relations

$$
\begin{aligned}
\text{all data points} \qquad & a \approx 0.0170\, b + 1.39\, l, \\
\text{leaving away } l > 0.1 \qquad & a \approx 0.0020\, b + 1.84\, l, \\
\text{leaving away } l > 0.05 \qquad & a \approx 0.0004\, b + 1.95\, l, \\
\text{leaving away } l > 0.025 \qquad & a \approx 0.0005\, b + 1.94\, l, \\
\text{leaving away } l > 0.0125 \qquad & a \approx 0.0002\, b + 1.98\, l.
\end{aligned}
$$

The resulting proportionality factor between a and l is closer to the one obtained for the stretch and shear specimens, but still slightly smaller. This difference might result from the coarse approximation in using expressions derived for a uniaxial stretch specimen. However, it confirms the finding for the stretch and shear specimens, that a precise value of an effective initial crack length cannot be introduced using the length scale parameter l. Further, the behavior of the phase field is again not representing brittle fracture, as can be seen by the non-local damage along the tensile fiber. This is also the case for the smallest length scale parameter chosen in the simulations, i.e. $l = 0.00625\,b$. For this choice of parameters, the ratio of the length scale parameter l and the characteristic size of the specimen size, i.e. the width b, is already below 1%. With respect to this ratio, the approximation of brittle fracture through the phase field method is still rather poor. Application of the phase field method to arrangements without high stress concentrators should therefore be treated with care.

5.3 Conclusion

A study on the role of the length scale parameter l as a material parameter is examined. Different specimen geometries are investigated and a principle correlation between l and an effective initial crack size a is found. However, the precise dependence of the crack length a on the length scale parameter l is found to depend on the specific geometry and boundary conditions. Further, the failure behavior does not resemble brittle fracture, but incorporates a damage-type process in advance to the actual formation of a crack.

We therefore do not recommend to use the length scale parameter in a practical way to adjust a certain crack length or material strength for general applications. Indeed, in coupled systems, where further physical parameters depend on the precise strain or stress of a purely elastic material, the damaging character may lead to erroneous results. Hence, in order to apply the method to situations, where no initial crack is introduced in advance, further development is mandatory.

The damage-type regime could, for example, be reduced by a modification of the prefactor, that releases the tensile energy, so that a significant increase of the order parameter d occurs only at a later stage, but then more abruptly. An alternative approach could involve an intermediate processing step, which follows a different criterion such as, for example, an evaluation of the stress in a certain volume. In this intermediate step, the order parameter could be set above a critical value. The subsequent crack growth could then again follow the energy based criterion.

6 Crack Growth during Lithium Insertion

In the previous chapters, it was shown that the phase field approach for fracture is a suitable method to generate reliable predictions of the appearance of crack growth under specific load conditions. Here, the fully coupled system of Li diffusion, mechanical stress and fracture is investigated during insertion of Li in a storage particle. The mechanical stress leading to crack growth is thus no longer caused by an externally prescribed load, but results from the distribution of Li in the material. At the same time, Li diffusion is strongly affected by the mechanical stress, in particular at the crack tip. The simultaneous study of all these physical processes reveals interesting phenomena, which are examined in the following.

The model incorporates a set of 8 material parameters. In addition, required inputs are the size of the storage particle, the magnitude of the applied flux and the length of a possible initial crack. Using the Buckingham π-theorem, this yields already 7 dimensionless groups [132], which could be investigated on their impact on fracture and diffusion in the storage particles. A comprehensive study on a sub-set of these groups without the introduction of a crack can be found in [167]. In order to reduce the complexity of the study and to obtain results, that are as general as possible, we focus on spherical (3D) or cylindrical (2D plane strain) particles[1] of outer radius r, as illustrated in FIG. 6.1. The boundary condition for the Li diffusion problem is applied in two steps. First, a constant flux $\bar{J}_0$ is imposed on the perimeter until the maximum concentration $c_{\max}$ is reached. Thereafter the concentration

[1] Real $LiMn_2O_4$ storage particles are often shaped like polyhedrons, with clear facets resulting from the crystallographic planes. However, the additional complexity, introduced by the orientation and size of the different surfaces, in particular with respect to the ones of a crack, would make an interpretation of the results in the sense of general conclusions harder.

c is held fixed[2]. The magnitude of the imposed flux $\bar{J}_0$ is prescribed in form of the C-rate C, i.e. the inverse time in hours theoretically required to completely charge or discharge the particle, through

$$|\bar{J}_0| = \frac{V}{S} c_{\max} C/[3600\,\mathrm{s}]. \tag{6.1}$$

Here, V and S are the storage particle volume and surface respectively. On battery cell level, this sequence corresponds to the so-called constant current constant voltage (CCCV) operation mode. In this mode the cell voltage is held fixed after a certain upper or lower cut-off voltage is reached during charging or discharging with a constant current. Under neglect of cell internal resistances present during operation, this is equivalent to keeping the Li concentration on the surface of the storage particles constant. The process of filling battery particles at a constant rate is also known as galvanostatic charging, while filling them at a constant voltage is known as potentiostatic charging.

For insertion, the largest tensile stress occurs in the interior of a particle, as shown, for example, in [129]. This stress might lead to fracture starting from voids or pores in the material, as found in [168]. An initial crack is therefore introduced in the particle center with the shape of a line (2D) or disk (3D) with length or diameter a. The material parameters are chosen by using values for $LiMn_2O_4$, as given in TAB. 6.1. The temperature is set to room temperature and for the critical energy release rate, a typical value of $10\,\mathrm{J/m^2}$ for a ceramic is assumed, which is in the range of experimental results for this material [169]. As in section 3.3.4, the characteristic length scale of the storage particles is defined by $r_0 = 1\,\mu\mathrm{m}$.

Following the findings in chapter 4, the initial crack is introduced as a heterogeneity in the phase field through an initially non-zero history field $\mathcal{H}_0$, as given by

$$\mathcal{H}_0(x, y) = \alpha\, e^{-(y/\beta)^2} \times \begin{cases} 1, & s < 0 \\ e^{-(s/\beta)^2}, & s \geq 0, \end{cases} \tag{6.2}$$

[2] The described scheme for Li insertion is numerically realized by a smoothed step-like function $\bar{J}(c)$, which takes on the value $\bar{J}_0$ for $c < c_{\max}$ and zero for $c \geq c_{\max}$. Hence, as soon as the maximum concentration $c_{\max}$ is reached at the perimeter of the storage particle, the concentration c oscillates around this value. However, the amplitude of the oscillation is negligible due to the comparatively small time steps in the simulations and does not affect the results obtained in this study.

where $\alpha = 10^4$ and $\beta = l/10$. The local distance to the crack tip is defined by $s = |x| - a/2$ in two dimensions and $s = \sqrt{x^2 + y^2} - a/2$ in three dimensions. To save computation time, we exploit the symmetry of the problem and simulate only one quarter of the particle in two dimensions, and one half of it in three dimensions. The simulations are started with zero concentration, i.e. $c_0 = 0$, and stopped when the state of charge (SOC), describing the mean concentration in units of maximum concentration $c/c_{\max}$, reaches a level above 99.9%.

Parameters				
Symbol	Description	Value	Unit	Ref.
E	Young's modulus	93	GPa	[168]
ν	Poisson ratio	0.3	-	[168]
ρ	material density	4140	kg/m^3	[170]
$c_{\max}$	max. concentration of Li	$2.29 * 10^4$	mol/m^3	[67]
Ω	partial molar volume of Li	$3.497 * 10^{-6}$	m^3/mol	[67]
D_0	diffusion coefficient of Li	$7.08 * 10^{-15}$	m^2/s	[66]
θ	temperature	300	K	
G_c	critical energy release rate	10	J/m^2	

Table 6.1.: Material parameters for lithium manganese oxide $LiMn_2O_4$.

6.1 Cylindrical Particles in 2D under Plane Strain

Particle sizes are chosen between $r = 5\,\mu m$ and $20\,\mu m$ in steps of $5\,\mu m$, the crack sizes are $a = 0.2\,\mu m$, $1\,\mu m$ and $2\,\mu m$ and the C-rate is given values of $C = 1, 5$ and 10. In all the simulations, the length scale parameter is set to $l = 0.05\,\mu m$. In this way, the range of transition between cracked and uncracked state is sufficiently small in comparison to the size of the crack. Hence, all crack sizes chosen are resolved correctly, including the smallest one. The residual stiffness parameter is chosen as $k = 10^{-5}$ and the mesh adaptation scheme is configured such that the mesh element size close to the crack faces has dimensions $h \leq (2/5)\,l$. Depending on the storage particle radius, the initial mesh and the stage of growth of the crack, the number of elements ranges from 828 to 80 397, or in terms of degrees of freedom the model ranges from 17 355 to 1 649 820. The computations are typically run on either 16 or 32 processors and take between 3 and 170 hours, heavily influenced by system size and crack behavior.

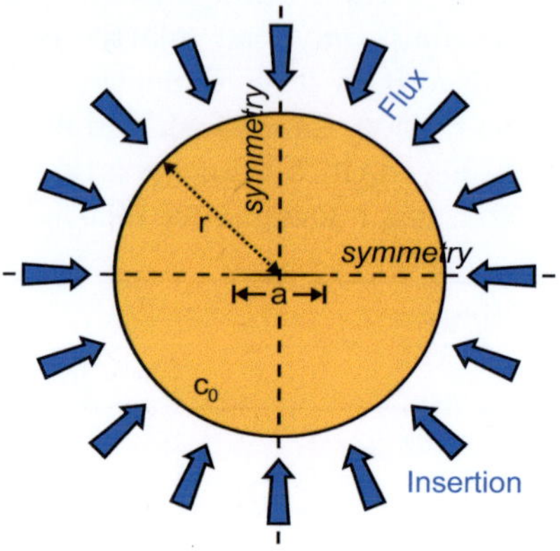

Figure 6.1.: Geometry and boundary conditions for the Li intercalation study. The geometry is either 2D plane strain in section 6.1 or 3D axisymmetric in section 6.2. In the three dimensional geometry, the initial crack has the shape of a disc. This is illustrated, for example, in FIG. 6.19.

As shown in TAB. 6.2, a number of different crack phenomena occur, depending the combination of parameters chosen. To present these in a comprehensible way, we begin with the condition with lowest mechanical stress in the storage particle and then go through the several possibilities of crack growth step by step.

6.1.1 Safe Conditions without Crack Growth

The least severe condition for the particle is given when its radius is smallest and the C-rate is low. This was also reported, for example, in [167] and results from the following argument. The time for Li to diffuse from the surface to the center of a storage particle grows approximately in a quadratic way with the radius of the particle. Consequently, a larger difference between the Li concentration at the perimeter and interior of the particle builds up when the radius is increased. This effect is further enhanced when the amount of Li inserted at the perimeter per unit time is high. The concentration difference leads to a non-homogeneous swelling of the particle and with it to mechanical stress, eventually resulting in fracture. The least harsh situation of the ones described in

TAB. 6.2 is therefore found when **r=5 μm**, **a=0.2 μm** and **C=1**. To understand the mechanism of stress generation, we take a closer look at the results of this simulation in the following.

Parameters			crack behavior		
$r\,[\mu m]$	$a\,[\mu m]$	C	growth	branching	breakage
5...20	0.2	1...10	no	no	no
5	1	1	no	no	no
5	1	5	yes	no	yes
5	1	10	yes	no	yes
10	1	1	yes	no	yes
10	1	5	yes	yes	yes
10	1	10	yes	yes	yes
15...20	1	1...10	yes	yes	yes
5	2	1	no	no	no
5	2	5	yes	no	no
5	2	10	yes	no	no
10...15	2	1...10	yes	no	no
20	2	1...10	yes	yes	yes

Table 6.2.: Input parameters, i.e. particle radius r, crack length a and C-rate C, as well as the occurrence of the different crack phenomena in the insertion study. Dots between two numbers imply that the corresponding line is valid for the entire range in between, e.g. storage particles with radius r between 5 and 20 μm and initial crack length $a = 0.2\,\mu$m neither experience crack growth, nor branching or breaking, when Li is inserted with a C-rate between 1 and 10.

Typical results for the process of Li insertion are shown in FIG. 6.2 a), where we use the term state of charge (SOC) to denote the average Li concentration in units of $c_{\max}$. The SOC first rises in a linear way to around 90% until its rate decreases at the transition from galvanostatic to potentiostatic charging at approximately $t \approx 23\,\tau$, with $\tau = r_0^2/D_0$. It then asymptotically converges to unity. The transition is marked by a sudden drop of the maximum concentration difference $\max \Delta c$, i.e. the difference between highest and lowest concentration in the particle.

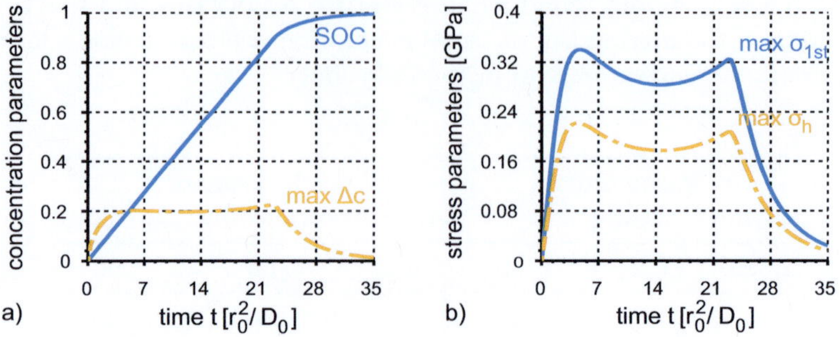

Figure 6.2.: Characteristic results for Li insertion with $r = 5\,\mu$m, $a = 0.2\,\mu$m and $C = 1$. (a) State of charge (SOC) and maximum concentration difference $\max \Delta c$ in units of maximum concentration $c_{\max}$. (b) The highest value of the maximum principal stress $\max \sigma_{1\text{st}}$ and maximum hydrostatic stress $\max \sigma_h$ during Li intercalation.

The role of the concentration difference on the generation of mechanical stress can be observed by comparing the maximum concentration difference $\max \Delta c$ in FIG. 6.2 a) with the highest value of the maximum principal stress $\max \sigma_{1\text{st}}$ in FIG. 6.2 b). The maximum concentration difference represents to some extent the magnitude of gradients of the Li concentration in the particle. As the maximum concentration difference grows to $\max \Delta c \approx 0.2\, c_{\max}$ in the first stage of the insertion process, $\max \sigma_{1\text{st}}$ increases also to around 0.34 GPa. This stress is caused by the higher Li concentration close to the particle surface compared to that at its center, as depicted in FIG. 6.3. The outer region of the particle expands more than the elastic stretching in the interior. Due to this process, tensile radial stress develops in the particle interior and compressive hoop stress results at its outer perimeter, as described, for example, in [129]. Both the highest value of the maximum principal and hydrostatic stresses are located at the crack tips.

After the concentration difference reaches its first maximum, it decreases slightly thereafter, beginning at $t \approx 4.9\tau$, before increasing again and peaking once more at the transition between galvanostatic and potentiostatic insertion. This small decrease results from the concentration dependent prefactor in the mechanical driving force of the Li flux (2.20), which is highest at $c = c_{\max}/2$ and thus hastens Li flux from the particle perimeter to its center as the concentration passes through this value.

However, as c approaches $c_{\max}$ near the perimeter, the diffusion there slows down again allowing a higher concentration difference build-up, causing a new peak in stress. Once galvanostatic insertion has ended, the Li distribution begins to equalize throughout the particle and the maximum concentration difference decreases, thereby relaxing the maximum stress.

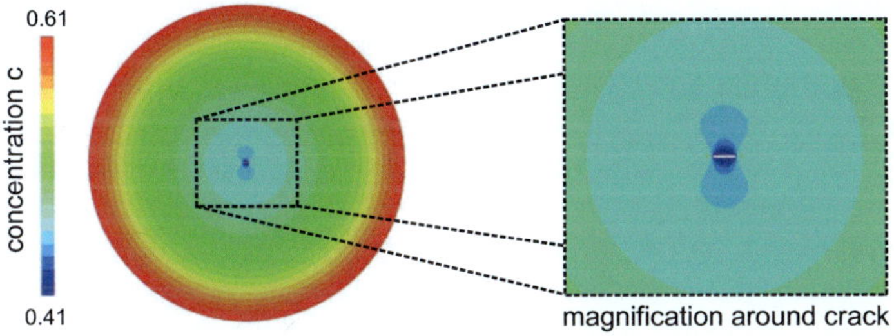

Figure 6.3.: Concentration c in units of maximum concentration $c_{\max}$ in a storage particle at an intermediate time step during Li insertion with $r = 5\,\mu\text{m}$, $a = 0.2\,\mu\text{m}$ and $C = 1$. The region around the crack is magnified on the right to show the impact of the stress around the crack on the concentration. Areas with $d > 0.95$ are removed to show the location of the crack. The opening of the crack is exaggerated by a linear scaling of the nodal displacements.

The graph of the highest value of the maximum principal stress $\max \sigma_{1\text{st}}$ versus time is therefore directly related to the time history of the maximum concentration difference $\max \Delta c$. Both quantities show peaks at the same point in time. However, two differences can be observed in FIG. 6.2. First, the relative decay of $\max \sigma_{1\text{st}}$ after the first peak is greater than that of the $\max \Delta c$ difference and second, $\max \sigma_{1\text{st}}$ is largest at the first peak, while $\max \Delta c$ is higher at the second one. We attribute these details to the effect of the crack on the stress distribution and its influence on the concentration.

The hydrostatic stress in the particle is shown in FIG. 6.4 for an intermediate time step in a simulation. As shown on the right inset of FIG. 6.4, the highest tensile hydrostatic stress is located at the crack tips. The same applies to the maximum principal stress. However, along the crack surfaces there is almost no stress at all ($|\sigma_h| < 1\,\text{MPa}$). As a consequence, the gradient of the hydrostatic stress, and with it the mechanical part of the flux, points towards the crack tips, but away from the crack faces. At some distance away from the crack ($\Delta r \approx 2...4\,a$), the stress field is to a good approximation spherically symmetric and only marginally influenced by the crack. Hence, the hydrostatic stress aids Li transport towards the particle center throughout most of the particle, but hampers Li diffusion to the crack faces at small distances from the crack, as illustrated on the right of FIG. 6.3. The stress gradient effect on Li flux is weighted by the prefactor $c(c - c_{\max}/2)$, see equation (2.20). At early stages of insertion, this prefactor is larger in regions away from the crack. The mechanical term in the Li flux therefore predominantly promotes diffusion into the particle center at the beginning of insertion. At later stages, the prefactor is higher close to the crack and stress-driven diffusion hinders transport to the particle center. The maximum concentration difference is therefore more pronounced at the later peak. However, while the concentration difference is slightly bigger at this point in time, the major part of the difference takes place over a small distance around the crack. For example, the relative difference in the concentration from the center to a distance $2\,\mu$m vertical above the crack in FIG. 6.3, is higher by 53% for the second peak, compared to the first one. This means, that the region with the lowest swelling is restricted to a smaller volume in the case of the later peak. The absolute difference in volume expansion between the outer and inner part of the particle is therefore smaller and less tensile stress is generated at the crack tip. Hence, the highest value of the maximum principal stress is larger at the first peak.

As described in section 2.3, the Li diffusion is not directly influenced by the crack, but only through its modification of the stress field. While this assumption may seem rather coarse at first glance, it is very reasonable for the examples presented here because of two reasons. First, a crack, lying completely in the interior of a particle, is not accessed by electrolyte. Hence, an additional sink or source of Li at the crack faces cannot be expected. And second, due to the symmetry of the examples considered, a driving force of Li through the crack, does not exist. It is therefore without effect whether the crack is modeled as permeable for Li diffusion or not. Other approximations associated with this assumption are believed to be negligible, for example the effect of surface diffusion of Li along the crack face.

The conditions lead to an example, where Li insertion is both limited by the rate of diffusion and the externally applied flux $\bar{J}_0$. A considerable concentration difference builds up, caused by the limited diffusivity and leading to a non-negligible stress in the particle. However, the diffusion of Li towards the center of the storage particle is still high enough to maintain an approximately constant concentration difference between its perimeter and interior during galvanostatic insertion. When larger particles or higher C-rates are considered, the effect of finite Li diffusion rate becomes more dominant. This can be seen in FIG. 6.5, where we plot the same concentration and stress parameters as before for the case of **r=5 μm**, **a=0.2 μm** and **C=5**, i.e. with a much higher C-rate.

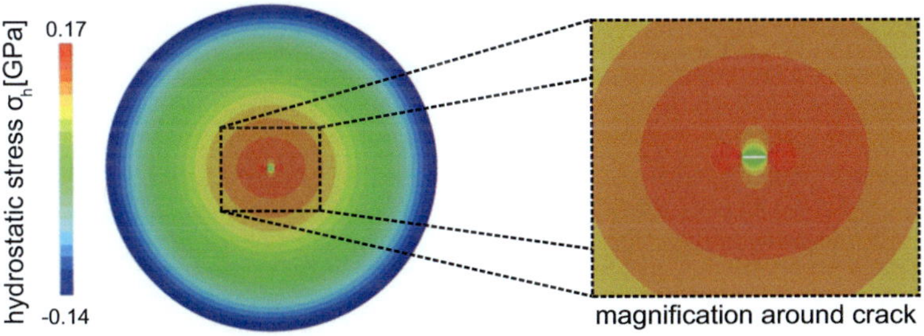

Figure 6.4.: Hydrostatic stress σ_h in a storage particle at an intermediate time step during Li insertion with $r = 5\,\mu\text{m}$, $a = 0.2\,\mu\text{m}$ and $C = 1$. The region around the crack is magnified on the right to illustrate the influence of the crack on the stress field. Areas with $d > 0.95$ are removed to show the location of the crack. The opening of the crack is exaggerated by a linear scaling of the nodal displacements.

The linear stage of the SOC graph now ends at $t = 3\,\tau$ with a value of SOC = 59%. The maximum concentration difference at this point is already at $\max \Delta c = 0.91\, c_{\max}$. This means, that while the particle is fully filled at its surface, there is only a very small amount of Li that has reached its center due to the slow diffusion. This leads to a noticeably higher maximum principal stress of around $\max \sigma_{1\text{st}} \approx 1.26\,\text{GPa}$, happening at the transition from galvanostatic to potentiostatic insertion. The maximum stress reaches even higher values when bigger particles or faster C-rates are considered. However, we find that the maximum value of $\sigma_{1\text{st}}$ saturates at around 1.48 GPa as the particle size and C-rate are increased, see TAB. 6.3.

The reason for the saturation can be understood through the following arguments. The typical time that Li needs to move a distance $\Delta r'$ in the radial direction is approximately $\tau_{\Delta r'} = \Delta r'^2 / D_0$. If the flux at the boundary is high enough and the particle big enough, the corresponding galvanostatic insertion time τ_{galvano} to reach the maximum concentration at the particle surface is much smaller then the typical time for the Li to diffuse into the particle, $\tau_{\text{galvano}} \ll \tau_{\Delta r'}$. The distance that Li travels by τ_{galvano} is negligible, $\Delta r' \ll r$. The insertion process is therefore almost entirely potentiostatic.

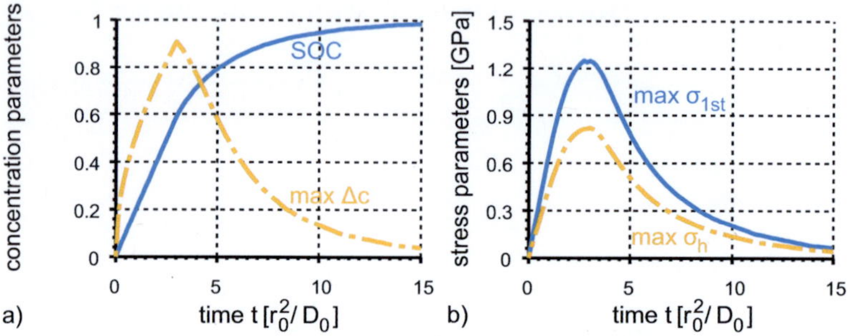

Figure 6.5.: Characteristic results for Li insertion with $r = 5\,\mu\text{m}$, $a = 0.2\,\mu\text{m}$ and $C = 5$. (a) State of charge (SOC) and maximum concentration difference $\max \Delta c$ in units of maximum concentration $c_{\max}$. (b) The highest value of the maximum principal stress $\max \sigma_{1\text{st}}$ and maximum hydrostatic stress $\max \sigma_h$ during Li intercalation.

We further note that for small crack lengths $a \ll r$, the modification of the stress distribution by the crack is constrained to a small region. Hence, in most of the particle, the pattern of Li diffusion is not significantly different from that in a particle without a crack. However, high stress in the central region of the particle does not result from small deviations in the Li content around the crack, but instead from large volume expansions in the rest of it. We may thus neglect the presence of a crack for the moment. Since the comparatively slow diffusion process is the only transient effect in this case, we may also neglect inertia terms, i.e. drop the left hand side of the conservation of linear momentum in equation (2.7). Under these assumptions, it can be shown through non-dimensionalization[3] of the corresponding constitutive and balance

[3] From the non-dimensionalized equations (A.2) and (A.7) in Appendix A.1, one can see that the balance equations of the boundary value problem do not depend on the radius r of the storage particle. The radius only scales the relevant equations by a factor. This becomes most obvious after application of the substitution rules (A.1), (A.4), (A.5) and (A.8), using the radius r as the characteristic

equations, that mechanical stress does not depend on the storage particle radius r, when Li is inserted potentiostatically. This was also shown by Cheng et al. [68] through analytic solutions for a spherical particle under neglect of stress-driven diffusion. In particular, equation (3) of their work states that the radial normal stress $\sigma_r(r') \propto c_{\mathrm{av}}^{\mathrm{total}} - c_{\mathrm{av}}(r')$ is proportional to the average concentration c_{av} in a spherical volume of radius r' within a particle with total average concentration $c_{\mathrm{av}}^{\mathrm{total}}$. The relation was taken from a thermal analogy in [171]. The conclusion remains the same for a cylindrical particle under plane strain.

Parameters $r\,[\mu\mathrm{m}]$	C	$\max\sigma_{1\mathrm{st}}$[GPa]	Parameters $r\,[\mu\mathrm{m}]$	C	$\max\sigma_{1\mathrm{st}}$[GPa]
5	1	0.340	15	1	1.443
5	5	1.246	15	5	1.463
5	10	1.456	15	10	1.463
10	1	1.097	20	1	1.476
10	5	1.477	20	5	1.478
10	10	1.477	20	10	1.478

Table 6.3.: Maximum values of the first principal stress $\sigma_{1\mathrm{st}}$ obtained for a crack length of $a = 0.2\,\mu\mathrm{m}$ and different particle radii and C-rates. The small difference between the magnitudes for $r = 15\,\mu\mathrm{m}$ and $r = 10\,\mu\mathrm{m}$ or $r = 20\,\mu\mathrm{m}$ at the higher C-rates is caused by different meshes.

The highest value of the maximum principal stress at higher C-rates in larger particles therefore only depends on the time τ_{galvano} to reach potentiostatic boundary conditions. As τ_{galvano} approaches zero, the highest value of the maximum principal stress remains constant. Since the crack is situated at the location of highest stress, the crack length $a = 0.2\,\mu\mathrm{m}$, which does not propagate in the simulations, thus is too small for Li insertion to lead to fracture through concentration difference induced stress, as shown in TAB. 6.2.

length r_0. The radius only comes into play through the boundary conditions, e.g. when applying pressure or a certain flux. This is shown by the substitution rule of the applied flux (A.3), which depends linearly on r_0. This dependence becomes quadratic when expressing the reference flux $\bar{\mathbf{J}}$ in terms of the C-rate. However, since the storage particle is free to expand and since in potentiostatic conditions only the Li concentration at the perimeter of the storage particle is set as boundary condition, the full boundary value problem is independent of r.

We find that the maximum stress generated in the particle depends on the ratio of two insertion limiting processes, the diffusivity of the Li-ions and the flux at the boundary. For large C-rates, the latter becomes negligible and the highest tensile stress occurs at the transition from galvanostatic to potentiostatic insertion. For a given particle radius, this stress determines the critical crack size still avoiding fracture.

6.1.2 Unstable Crack Growth

As described in the above section, high mechanical stress is expected for large C-rates. From Griffith's theory for brittle fracture it is further known that the critical stress for a crack to grow typically decreases with larger crack size. In order to simulate fracture in a storage particle, we therefore choose the parameter combination $\mathbf{r{=}5\,\mu m}$, $\mathbf{a{=}2\,\mu m}$ and $\mathbf{C{=}5}$, i.e. a high C-rate with a long crack. The associated phenomena discussed in the following are representative for a wider set of input parameters. As shown in TAB. 6.2, similar behavior is obtained for $C = 10$ and in all simulations with $a = 2.0\,\mu\text{m}$ and $10\,\mu\text{m} \leq r \leq 15\,\mu\text{m}$.

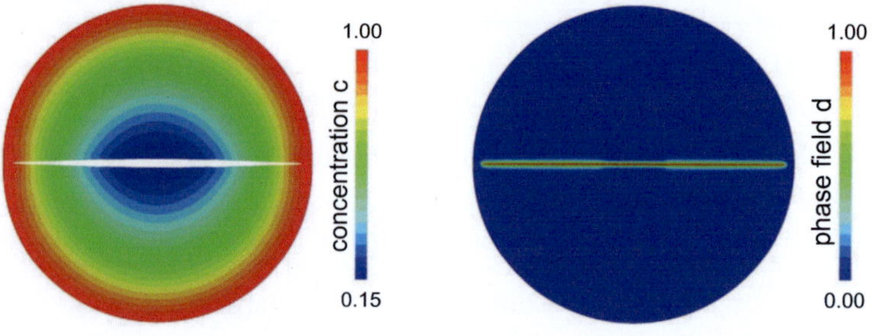

Figure 6.6.: (Left) Concentration c in units of maximum concentration $c_{\max}$ in a storage particle with $r = 5\,\mu\text{m}$, $a = 2\,\mu\text{m}$ and $C = 1$ after crack growth has stopped. Areas with $d > 0.95$ are removed to show the location of the crack. The opening of the crack is exaggerated by a linear scaling of the nodal displacements. The spherical symmetry of the concentration distribution is perturbed due to the presence of the crack. (Right) Order parameter d at the same time step.

As the crack length is increased, the stress at the crack tip becomes large enough to initiate crack growth. Assuming that the stress at the crack in the interior results mainly from the concentration gradient close to the boundary, the stress field at the crack in the first instants of growth is to some extent comparable to that at a small through crack in a very large plane strain specimen subject to biaxial stress at its perimeter. In this case, crack propagation occurs in an unstable way since the stress intensity factor increases with crack length even if the applied load is held fixed. Of course, the stress field in the storage particle is more complicated and, with extension of the crack, the situation resembles more closely a displacement controlled process rather than a load controlled one as the extension of the crack will tend to relax the particle stresses. Above all, close to the surface of the particle the radial stress drops to zero, as mentioned above, and the hoop stress becomes compressive, as described in [129]. The crack thus grows into a compressive stress field limiting further propagation.

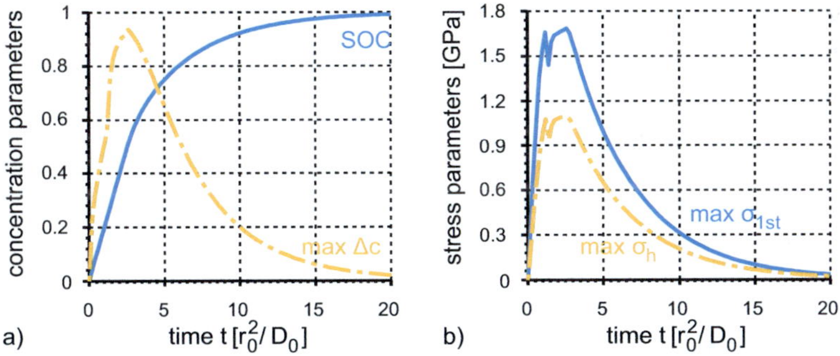

Figure 6.7.: Characteristic results for Li insertion with $r = 5\,\mu$m, $a = 2\,\mu$m and $C = 5$. (a) State of charge (SOC) and maximum concentration difference $\max \Delta c$ in units of maximum concentration c_{max}. (b) The highest value of the maximum principal stress $\max \sigma_{1\text{st}}$ and maximum hydrostatic stress $\max \sigma_h$ during Li intercalation.

As will be shown in the following, the principle idea of the crack growth behavior during Li insertion as described above is confirmed in the numerical simulations. The crack first penetrates in an unstable way and then comes to rest close to the surface, as shown in FIG. 6.6. The graphs of SOC and stress are plotted for this case in FIG. 6.7. We observe that the curve of the SOC follows the same trend as before. It rises first linearly and after the transition from galvano- to potentiostatic insertion its rate of increase falls. In contrast, the stress parameters behave significantly differently. They both first peak, drop and then rise to the same magnitude again. For the maximum principal stress this peak is

$\max \sigma_{1\mathrm{st}} = 1.68\,\mathrm{GPa}$. Also the maximum concentration difference shows a small deviation from the curves for cases without crack growth in the early stages of insertion. More detail is shown in FIG. 6.8.

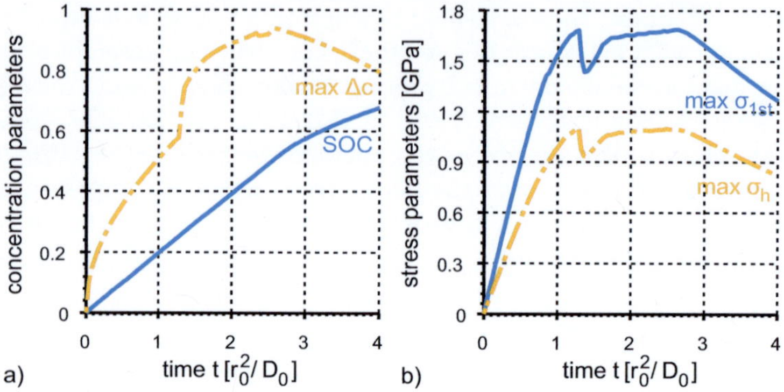

Figure 6.8.: Magnified view on characteristic results of FIG. 6.7 for early times. (a) State of charge (SOC) and maximum concentration difference $\max \Delta c$ in units of maximum concentration $c_{\max}$. (b) The highest value of the maximum principal stress $\max \sigma_{1\mathrm{st}}$ and maximum hydrostatic stress $\max \sigma_h$ during Li intercalation.

Comparing SOC and maximum first principal stress in FIG. 6.8, we find that the highest stress is reached before the boundary condition switches from galvanostatic to potentiostatic insertion at $t \approx 2.6\,\tau$. With respect to the above considerations regarding maximum stress and concentration difference, this indicates that the stress that can be generated with this set of parameters is higher than that necessary to initiate crack growth. Most of the crack growth takes place at $t \approx 1.31\,\tau$. At this time the crack starts to propagate in an unstable way on a time scale that is smaller by orders of magnitude than that associated with the diffusion process. This is shown by a step-like behavior of the graph of crack length versus time plotted in FIG. 6.9. For this reason, unstable crack growth is associated only with the first peak of the stress curve. After this peak, the maximum principal stress drops rapidly to a value below $\max \sigma_{1\mathrm{st}} = 1.5\,\mathrm{GPa}$.

We attribute this dramatic drop to inertia driven crack propagation in the last stages of growth. The material adjacent to the upper and lower crack faces build up momentum when the crack is rapidly propagated. The deceleration of the material then leads to an inertia induced stress field in addition to the stress from the volume expansion, when crack growth is slowed down, causing further crack growth. The additional stress field then relaxes on a time scale controlled by inertia effects, explaining the dramatic drop in the maximum stress. Thereafter the remaining stress is due only to the inhomogeneous volume expansion of the material associated with the Li distribution.

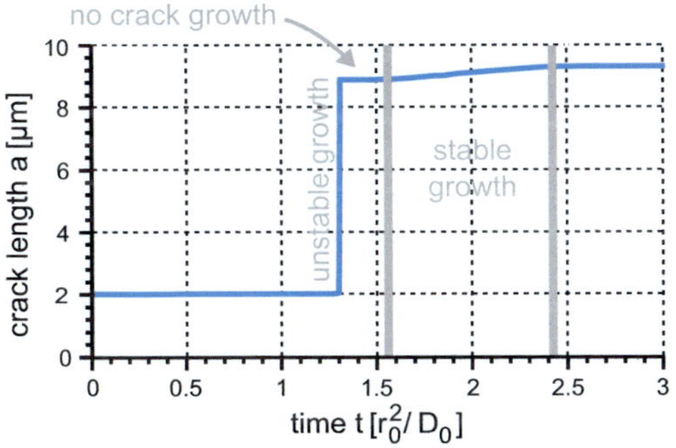

Figure 6.9.: Crack length during Li insertion in a storage particle with radius $r = 5\,\mu\text{m}$, initial crack length $a = 2\,\mu\text{m}$ and C-rate $C = 5$. Unstable crack growth is followed by an intermediate phase of arrested and subsequent stable crack growth.

Due to the small time scale of the unstable crack growth, the Li distribution is not able to adjust simultaneously to its stress field. Only after crack propagation has ended does Li redistribution become recognizable. Relative to the rate of Li diffusion, the crack tip propagates very rapidly to its penultimate location. The crack tip is surrounded by a region of high tensile stress, which propels the Li to flow towards the tip. This effect is shown in FIG. 6.10. The mechanically driven flux becomes large enough, so that the concentration at the crack tip exceeds the maximum concentration in the rest of the particle. This explains the abrupt change of slope and the subsequent elevation in the graph of the maximum concentration difference in FIG. 6.8. Hence, the location of highest concentration is no longer at the surface, but at the crack tip. The effect is so strong that the concentration on the surface closest to the crack tip is lower than on the rest of the perimeter.

As more Li flows into the particle, the magnitude of the stress driven diffusion decreases because of the concentration dependent prefactor. A front of almost maximum concentration then passes the crack tip and thereafter the Li concentration distribution has little dependence on the stress.

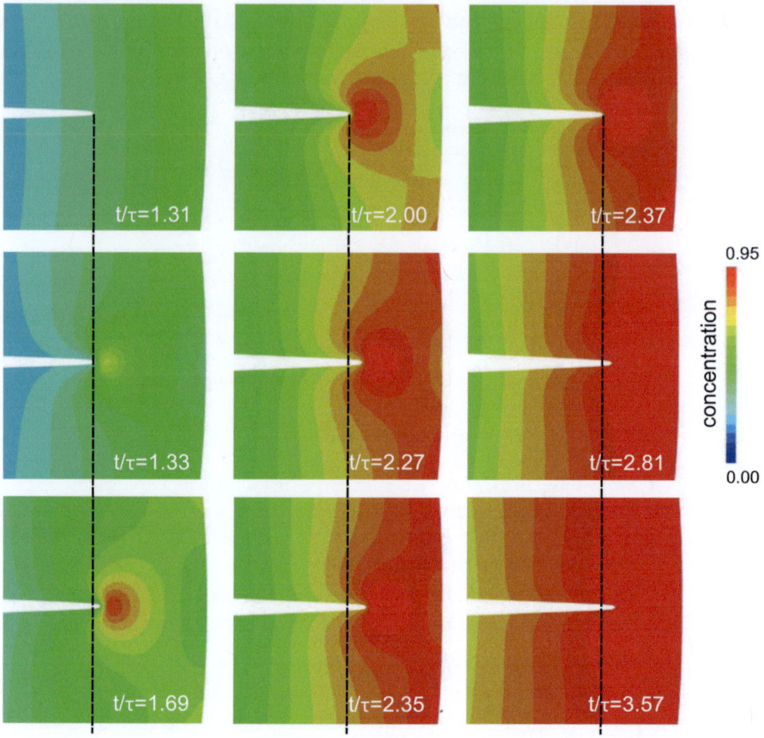

Figure 6.10.: Li concentration in units of maximum concentration $c_{\max}$ redistributing around the crack tip after unstable crack growth has stopped around $t/\tau \approx 1.3$ with $\tau = r_0^2/D_0$. Areas with $d > 0.95$ are removed to show the location of the crack. The opening of the crack is exaggerated by a linear scaling of the nodal displacements. Dashed lines are drawn to aid recognition of the extension of the crack during stable growth. The illustrations are obtained from a simulation with $r = 5\,\mu$m, $a = 2\,\mu$m and $C = 5$. As unstable crack propagation is rapid, Li only starts to adapt to the new stress field after growth has stopped. This can be seen in the top left, where lines of equal concentration are still parallel to the surface.

The Li inserted into the particle, after the unstable crack propagation has ended, leads again to an increase of the stress in the particle and in particular at the crack tip. This corresponds to the short, steep increase after the drop in the maximum principal stress of FIG. 6.8. During the build-up of stress the crack does not grow further. This is shown by the short, constant plateau between unstable and stable crack growth in FIG. 6.9. The stress at the crack tip then becomes large enough, $\sigma_{1st} \approx 1.62\,\text{GPa}$, to drive the crack to grow a little further. This is marked by the noticeable decrease in the slope of the maximum principal stress at around $t \approx 1.69\,\tau$.

The crack then propagates a small distance in a stable way, as can be observed in FIG. 6.9. The stage of stable crack growth thereby corresponds to the segment of smaller slope in σ_{1st}, shown in FIG. 6.8, that ends at $t \approx 2.62\,\tau$ with the maximum stress $\max \sigma_{1st} \approx 1.68\,\text{GPa}$. The time at which crack extension ceases coincides approximately with the transition from galvanostatic to potentiostatic insertion. However, we do not attempt to explain the criterion controlling crack growth termination in terms of global parameters, due to the complicated stress field resulting from the Li concentration around the crack, as shown in FIG. 6.10.

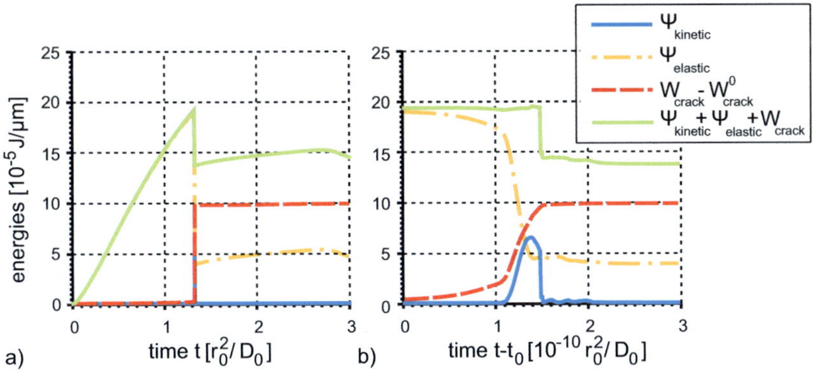

Figure 6.11.: (a) Kinetic, elastic and fracture energy, as well as their sum, around the moment of crack growth during Li insertion in storage particle with $r = 5\,\mu\text{m}$, $a = 2\,\mu\text{m}$ and $C = 5$. Since the initial crack is considered as a feature of the investigated geometry, the offset associated with its creation, W^0_{crack}, is subtracted from the fracture energy W_{crack}. (b) Detail of the diagram in (a) for the time during crack growth. The offset t_0 is set such that it approximately coincides with the initiation of crack growth.

The mechanism of momentum build-up, that drives the crack a little further than expected in a static situation, can also be understood by means of FIG. 6.11, where we show the different energy contributions related to crack propagation. The elastic energy Ψ_{elastic} in the particle first rises due to the insertion of Li and the resulting inhomogeneous concentration distribution. It is then released at the onset of crack growth. At the same time, some of it is transformed into kinetic energy Ψ_{kinetic} due to movement of separation of the material adjacent to the crack faces. The elastic energy than reaches a stationary value, comprised of both compressive and tensile contributions in the entire particle but it is is not localized enough to drive further crack growth. Next, kinetic energy is partly converted back into elastic energy, which in turn enables the creation of new fracture surfaces. This is reflected in the slightly differing times at which elastic energy Ψ_{elastic} and the final fracture energy W_{crack} reach a stationary value, as shown on the right of FIG. 6.11.

On the short time scale of unstable crack growth, the free energy contributions resulting from the amount and distribution of Li in the particle do not change significantly. Except for a constant offset, the summed contributions represented by the green curve in FIG. 6.11 are therefore equivalent to the total free energy during crack growth. With respect to the green curve, we therefore find that energy conservation does not hold, even on time scales for which Li insertion is negligible. This results from two reasons. First, the model describes isothermal processes, i.e. the temperature is kept constant and generated (dissipated) heat is absorbed (provided) by the environment. Second, although actual heat generation from fast motions in the body, such as oscillations or vibrations, is not modeled by physical means, it is qualitatively covered in the numerical implementation. When crack growth stops, the time steps in the numerical simulation increase by orders of magnitudes. Inertia effects cannot be resolved with these comparatively large time steps and the kinetic energy is neglected. Energy conservation is therefore only violated after the kinetic energy has been dissipated, which can be seen by the simultaneous drop of the blue and green curve on the right hand side of FIG. 6.11. Of course, this is not a suitable approach when truly modeling effects such as wave propagation in the material. However, in this case, the behavior is qualitatively equivalent to that expected if a damping term, modeling energy dissipation in the sense of heat generation, were introduced.

In order to explain the appearance of unstable crack growth by means of analytic considerations, we wish to compare the competing effects similar to the idea of an R-curve [132]. The following equations hereby result from very simplified assumptions and shall only serve for a qualitative understanding of the phenomenon.

The stress intensity factor in an infinite plane, i.e. under negligence of boundary effects, with a uni- or biaxial applied load of magnitude σ is given by $K_I = \sigma\sqrt{\pi a}$. It is related to the energy release rate G in plane strain through $K_I^2 = GE/(1-\nu^2)$. Combining both relations, we find the critical stress at which fracture occurs to be

$$\sigma_c(a) = \frac{1}{a^{1/2}}\sqrt{\frac{GE}{\pi(1-\nu^2)}}. \tag{6.3}$$

With extension of the crack length a, the critical stress decays as

$$\frac{\partial \sigma_c}{\partial a} = -\frac{1}{2}\frac{1}{a^{3/2}}\sqrt{\frac{GE}{\pi(1-\nu^2)}}. \tag{6.4}$$

A simple relation, expressing a condition for crack growth in terms of a characteristic stress σ_c, as given in equation (6.3), does not exist for a crack in a storage particle. However, to gain a qualitative understanding, we take as characteristic stress the radial stress σ_r in a particle without a crack[4]. The radial stress is found from a thermal analogy in [171] as

$$\sigma_r(r') = \frac{\Omega E}{6(1-\nu)}\left(c_{\mathrm{av}}(r) - c_{\mathrm{av}}(r')\right), \tag{6.5}$$

with the average concentration $c_{\mathrm{av}}(r') = \frac{2}{r'}\int_0^{r'} c(\tilde{r}')\tilde{r}'\mathrm{d}\tilde{r}'$ within the radius $r' \in [0, r]$ around the center of a particle with radius r. We assume a Li distribution along the radius that can be approximated by a power law $c(r) = (c_o - c_i)(r'/r)^\gamma + c_i$ with the inner and outer concentration c_i and c_o respectively. The Li content along the radius is typically convex, which requires an exponent of $\gamma > 1$. The radial stress then reads

$$\sigma_r(r') = \frac{\Omega E}{3(1-\nu)}\frac{c_o - c_i}{\gamma + 2}(1 - (r'/r)^\gamma). \tag{6.6}$$

To compare the decay of critical and radial stress, we choose the concentration difference $c_o - c_i$ such, that $\sigma_r(r') = \sigma_c(a)$. The derivative of σ_r with respect to the radius then becomes

[4] The same argument can also be carried out with the circumferential stress. However, while the trends remain the same, the resulting expressions become more complicated.

$$\frac{\partial \sigma_r}{\partial r'} = -\frac{1}{a^{1/2}} \frac{\gamma}{r^\gamma} \frac{r'^{\gamma-1}}{1-(r'/r)^\gamma} \sqrt{\frac{GE}{\pi(1-\nu^2)}}. \tag{6.7}$$

To obtain a rule of thumb on how the radial stress decreases with r' close to the crack tip, we set $r' = a$. The strongly modified stress field at the crack tip is neglected hereby. Unstable crack growth then occurs if $\partial\sigma_c/\partial a < \partial\sigma_r/\partial r'_{|r'=a}$. Using (6.4) and (6.7), this yields the stability condition

$$\frac{1}{2} \frac{\sqrt{r}}{\gamma} \frac{1-(a/r)^\gamma}{(a/r)^{\gamma-1/2}} < 1. \tag{6.8}$$

If the left hand side is larger then one, the crack will propagate in an unstable way. The derived relation significantly underestimates the range of crack lengths that undergo unstable propagation. However, it provides a number of fundamental dependencies. The function is monotonically increasing with γ, i.e. with the convexity of the Li distribution. Thus, as more Li is located at the outer parts of the particle, unstable crack growth is more likely to occur. The same applies for an increasing radius of the particle r when the ratio a/r is fixed. And lastly, as a approaches r, crack propagation is expected to be stable.

We have seen that a certain minimum length is required for the crack to grow even when particle size and C-rate are increased to high values. At the same time, if crack growth starts from a through crack in the center, it is most likely to happen in an initially unstable way, followed by a distinct, short period of stable growth.

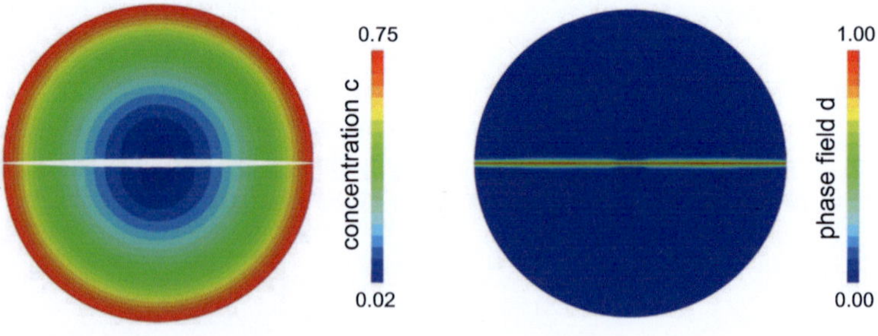

Figure 6.12.: (Left) Concentration c in units of maximum concentration c_{max} in a storage particle with $r = 5\,\mu$m, $a = 1\,\mu$m and $C = 5$ after it has broken into two. Areas with $d > 0.95$ are removed to show the location of the crack. The opening of the crack is exaggerated by a linear scaling of the nodal displacements. (Right) Order parameter d at the same time step.

6.1.3 Particle Breakage

We define the critical crack size a_c as the maximum length of the initial crack a that does not lead to crack growth during Li insertion in a storage particle of a given radius r during Li insertion with a defined C-rate C. In order inspect a crack length a that is close to the critical one, a_c, we decrease a and investigate the parameters **r=5 μm**, **a=1 μm** and **C=5**. This combination of parameters is also representative of the case of $C = 10$, as shown in TAB. 6.2.

As a result of the smaller initial crack length, the crack now propagates all the way through to the perimeter of the particle and breaks it into two. We emphasize that this happens in only one half cycle of lithiation. As can be seen in FIG. 6.12, the fracture process again takes place on a time scale short enough so that the Li concentration distribution is barely able to adapt to the stress field modified by the extended crack.

The reason for the breakage of the particle lies in the momentum build-up during crack growth, as already described in section 6.1.2. However, in comparison to the case studied in section 6.1.2 the effect is more pronounced due to the smaller initial crack length. To illustrate this point, the crack tip velocities for the cases with initial crack length $a = 1\,\mu$m and $a = 2\,\mu$m are plotted in FIG. 6.13. In order to extract the crack tip velocity from a sequence of contour plots, the crack tip is defined to be located at the position at which the order parameter takes on the value $d = 0.95$.

The offsets for the starting time of crack growth are shifted in FIG. 6.13 such that both graphs for the two initial crack lengths can be displayed in one diagram. The actual starting time for the smaller initial crack length ($t \approx 1.98\,\tau$) is significantly larger than for the larger initial crack length ($t \approx 1.29\,\tau$). This is due to the higher load necessary to initiate crack growth of a smaller crack. To generate this higher load through an inhomogeneous concentration of Li, more Li has to be inserted through the boundary and more time is required. As a consequence, more elastic energy is accumulated in the particle when the smaller initial crack begins to grow ($\psi_{\text{mech}}/\mu\text{m} = 6.5 * 10^{-5}\,\text{J}/\mu\text{m}$ in comparison to $\psi_{\text{mech}}/\mu\text{m} = 4.8 * 10^{-5}\,\text{J}/\mu\text{m}$). As this energy is released, the growth of the crack is significantly faster, as can be seen both by the maximum velocity and the comparatively short time to break the particle. The large elastic energy accelerates the material adjacent to the upper and lower faces of the crack faster away from each other.

Hence, more linear momentum builds up and the inertia induced additional stress, generated when crack growth slows down, is sufficient to drive the crack all the way to the surface.

Inertia effects, resulting from unstable crack growth, may lead to particle breakage in a single Li insertion half cycle. This phenomenon is most likely to occur when the initial crack length is close to the critical one.

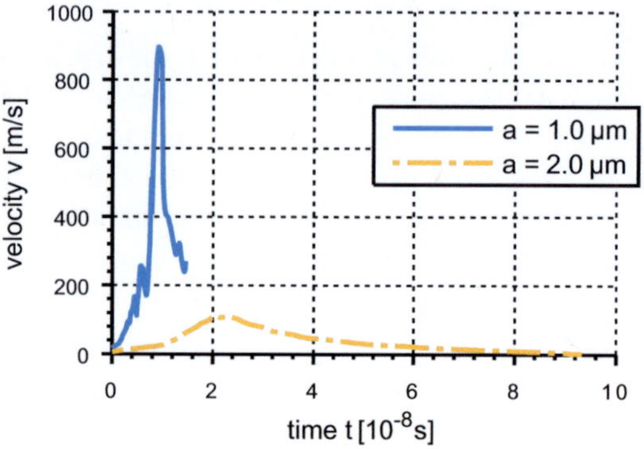

Figure 6.13.: Crack tip velocity against time for two different initial crack lengths. Particle radius and C-rate are $r = 5\,\mu$m and $C = 5$ in both cases. The starting time is shifted so that both cracks begin to propagate at $t = 0$ s.

6.1.4 Crack Branching

Inertia driven phenomena become more distinct when the particle radius is increased and the crack length is kept at an intermediate value, i.e. above the critical one a_c, but still small enough so that a large amount of elastic energy is stored in the particle before crack growth is initiated. In order to study possible effects related to higher crack tip velocities than those observed above, we examine the combination $\mathbf{r{=}10\,\mu m}$, $\mathbf{a{=}1\,\mu m}$ and $\mathbf{C{=}5}$. With respect to the phenomena observed, the set of parameters is equivalent to the same combination with $C = 10$, see TAB. 6.2.
The larger particle radius now leads to crack branching, as plotted in FIG. 6.14. The reason for branching lies in the increased crack tip velocity. On the one hand, this results from the larger travel distance through regions of high tensile stress, during which the crack tip is fur-

ther accelerated. And on the other hand, the ratio of initial crack length and radius is smaller, so that the crack only starts to grow at a later stage, when more elastic energy is stored in the particle. This is also in agreement with the stability condition (6.8), which yields values that are further away from the condition for stable growth when the radius of the storage particle is larger and more Li is close to its perimeter, as obtained with higher C-rates.

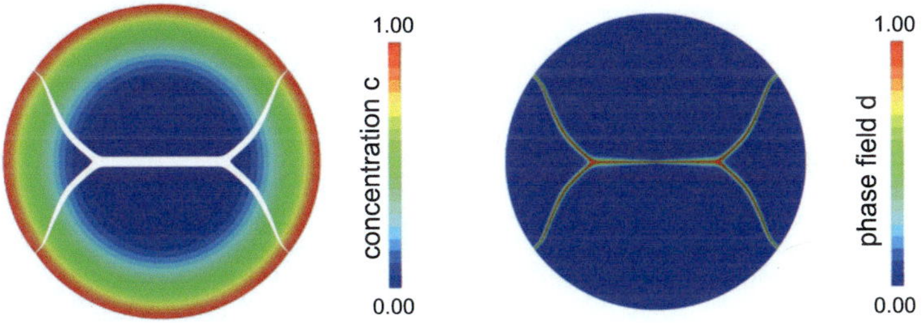

Figure 6.14.: (Left) Concentration c in units of maximum concentration c_{max} in a storage particle with $r = 10\,\mu\text{m}$, $a = 1\,\mu\text{m}$ and $C = 5$ after cracks have branched and reached the surface. Areas with $d > 0.95$ are removed to show the location of the crack. The opening of the crack is exaggerated by a linear scaling of the nodal displacements. (Right) Order parameter d at the same time step.

The higher crack tip velocity is also observable in FIG. 6.15, where we plot the crack tip velocities during Li insertion with two different C-rates. The precise point of branching cannot be determined for a finite value of the length scale parameter l, since the phase field first starts to broaden a little and then branches off. The corresponding time ranges are marked by gray circles in FIG. 6.15. The typical wave speeds for this system are, $c_1 = 5730\,\text{m/s}$, $c_2 = 3062\,\text{m/s}$ and $c_r = 2841\,\text{m/s}$ for the dilatational, shear and Rayleigh wave speed respectively [172]. The maximum crack tip velocity lies between 1400 and 1600 m/s, which is in a qualitatively satisfying agreement with, for example [173], where the critical velocity for crack branching is found to be $v_{\text{crit}} = 0.46\,c_r$. The value is taken from solutions of [174] for a mode III problem[5], but should also give a reasonable estimate for this case.

[5] It is derived using Griffith's energy release rate criterion [133, 175, 176] and the principle of local symmetry [177–179], which states that the crack propagates in such a direction that in-plane shear stress always vanishes in the vicinity of the crack tip.

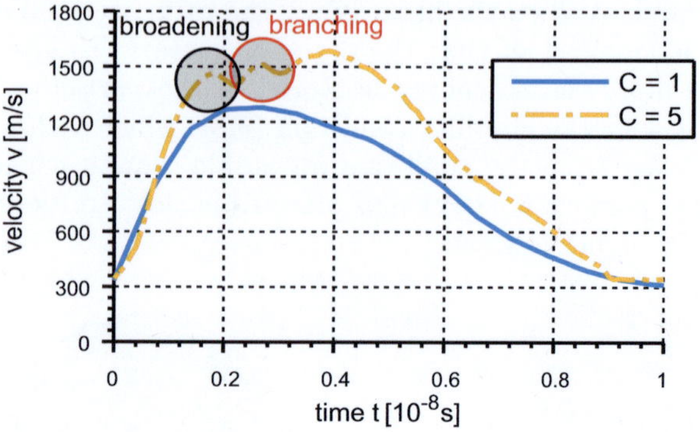

Figure 6.15.: Crack tip velocity against time for two different C-rates. Particle radius and crack length are $r = 10\,\mu$m and $a = 1\,\mu$m in both cases. The starting time is shifted so that both cracks begin to propagate at $t = 0$ s. The regions in gray mark times where the phase field broadens and branches.

The fundamental reason for crack branching lies in a velocity dependent modification of the stress distribution around the crack tip. In terms of the stress intensity factors, this means, that their dynamic counterparts $K_l(t, v)$ are related to the ones at rest $K_l(t, 0)$ [173, 175, 176, 180] through

$$K_l(t, v) = k_l(v) K_l(t, 0), \tag{6.9}$$

where v is the crack tip velocity, $k_l(v)$ are known universal functions of v and $l = \mathrm{I}, \mathrm{II}, \mathrm{III}$ denotes the respective mode. The modification of the stress around the crack tip during crack growth and branching is illustrated in FIG. 6.16 by means of the maximum principal stress. We observe, that with increasing crack tip speed, the region of high tensile stress is distorted more and more around the crack tip so that it is finally preferential for the crack tip to branch into two.

We find that such a strong distortion of the stress field is absent for slower crack growth. This is illustrated in FIG. 6.17, which again shows the maximum principal stress during crack growth, but taken from a simulation with parameters $r = 5\,\mu$m, $a = 1\,\mu$m and $C = 5$. Crack branching does not occur in this case. The corresponding condition for the critical velocity v_{crit}, at which a crack may branch, can be deduced from a balance of energy release, as is typical in Griffith's theory [133, 173]. Due to the modification of the stress field around a fast moving crack tip, the energy release rate becomes velocity dependent $G = G(v)$. The same applies for the energy release rate $\tilde{G}$ of a single moving crack after branching. However, the latter also depends on the branching angle ϕ, i.e. $\tilde{G} = \tilde{G}(\tilde{v}, \phi)$, where $\tilde{v}$ is the crack tip velocity after branching. As always, the energy release rate must be equal to the critical one, i.e. the energy to tear apart the crack faces, which in general may also depend on the velocity. Assuming that the fracture energy is constant, one can deduce a relation between the energy release rates immediately before and after branching, $G(v) = \tilde{G}(\tilde{v}, \phi)$ [173]. This imposes a necessary condition for crack branching. For small initial velocities v, the energy release rate of the single crack $G(v)$ is always larger than $G'(v', \phi)$ and branching cannot occur [173].

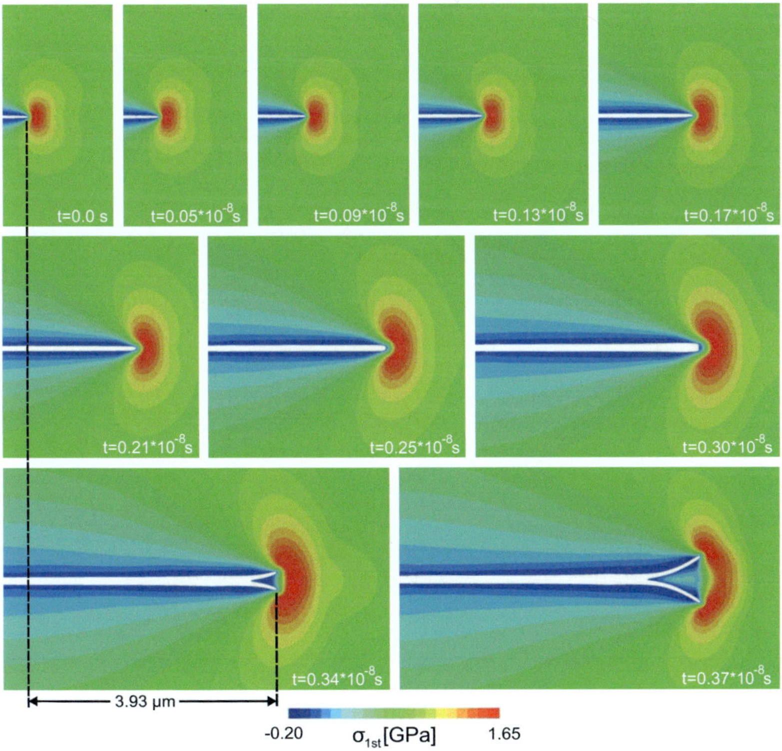

Figure 6.16.: Maximum principal stress around crack tip for various states of crack growth in the storage particle with $r = 10\,\mu$m, $a = 1\,\mu$m and $C = 5$. Areas with $d > 0.95$ are removed to show the location of the crack. The opening of the crack is exaggerated by a linear scaling of the nodal displacements.

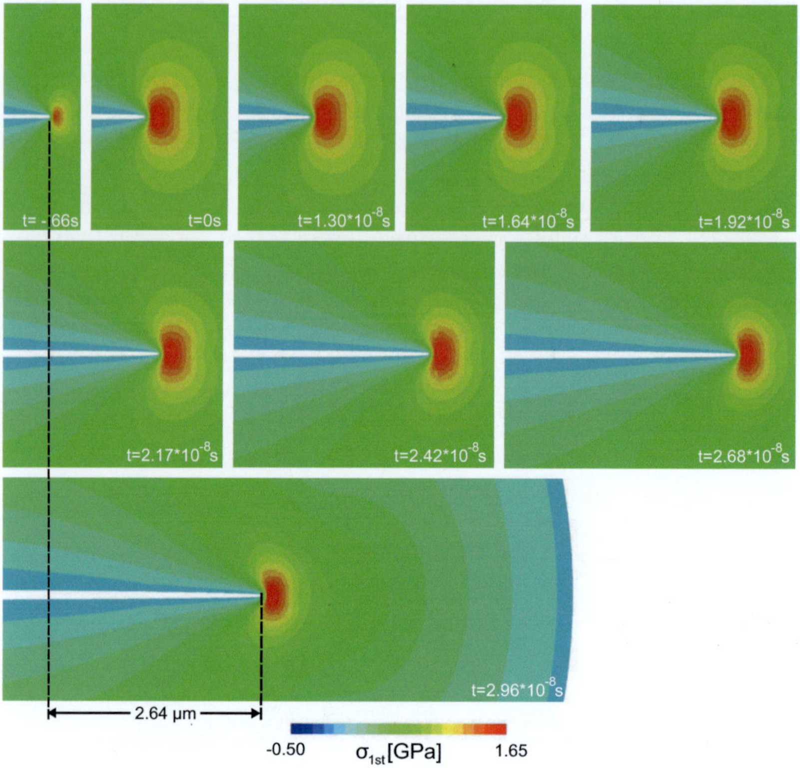

Figure 6.17.: Maximum principal stress around the crack tip in a storage particle with $r = 5\,\mu\text{m}$, $a = 1\,\mu\text{m}$ and $C = 5$. Areas with $d > 0.95$ are removed to show the location of the crack. The opening of the crack is exaggerated by a linear scaling of the nodal displacements. In contrast to the scenario with crack branching, the region of high stress is only slightly distorted around the crack tip.

Because of the crack branching, the path of the crack no longer lies on the symmetry axis between the upper and lower half of the particle. Hence, the argument given in section 6.1.1, that the condition for flux through the crack can be ignored due to symmetry is no longer valid. However, since crack growth takes place in an unstable way in these examples, diffusion is too slow to change the Li distribution significantly before crack propagation stops or before the crack reaches the surface.

Since we are only interested in the time range until breakage, neglecting a direct impact of the crack on diffusion is still a good approximation.

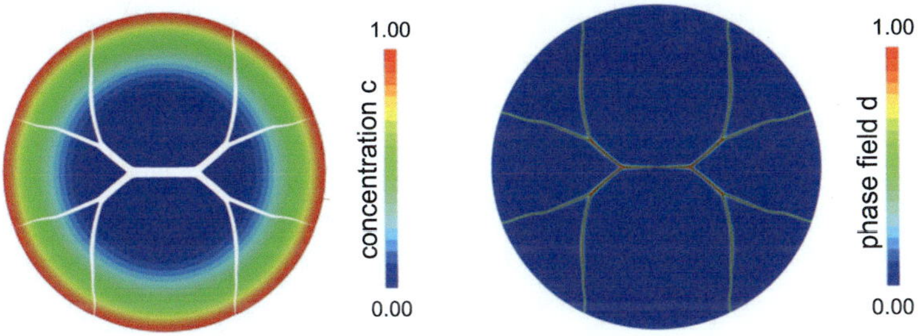

Figure 6.18.: (Left) Concentration c in units of maximum concentration c_{max} in a storage particle with $r = 20\,\mu\mathrm{m}$, $a = 1\,\mu\mathrm{m}$ and $C = 5$ after cracks have branched twice and reached the perimeter. Areas with $d > 0.95$ are removed to show the location of the crack. The opening of the crack is exaggerated by a linear scaling of the nodal displacements. (Right) Order parameter d at the same time step.

Going to even more severe conditions, a second branching occurs and the fracture patterns become increasingly complex, as depicted in FIG. 6.18. The combination of parameters used for this simulation is **r=20 μm, a=0.5 μm** and **C=5**. Similar crack topologies are also found for other C-rates. We observe that the second branching occurs in a region of almost zero concentration, i.e. at a location that has not yet been reached by a significant amount of Li. This circular area is marked by the deep blue color in the left plot of FIG. 6.18. In the considerations on stress in a particle without crack growth in section 6.1.1, it is shown that the stress at a certain radial distance from the center of a storage particle is to a great extent determined by the average Li concentrations of the material inside and outside of this radius. This means, that the stress in the interior of the deep blue zone merely decreases with distance to the center. Hence, after the first crack branches and the new crack tips are thus moving relatively slowly, the crack tips are again accelerated by high stress to the critical velocity v_c for crack branching.

Due to inertia effects the particle may therefore not only break into two, but be fragmented into several parts. This phenomenon is an outcome of crack branching, occurring if the crack tip velocity exceeds a critical value v_c, achieved in cases of large particle radii r and C-rates C and with initial crack lengths a close to the critical one a_c.

6.2 Spherical Particles in 3D

The two dimensional simulations have revealed different effects for crack growth and its interplay with Li diffusion. To confirm that these findings are transferable to three dimensional storage particles, we run a small number of computations in 3D. Due to the computational cost of these simulations, only three combinations of parameters are examined. As shown in TAB. 6.4, they are chosen such that the different fracture phenomena, described in section 6.1, i.e. crack growth, branching and breakage, are represented. Further effects resulting from the additional dimension shall be discussed in the following.

Parameters			crack behavior		
r [μm]	a [μm]	C	growth	branching	breakage
5	1	5	no	no	no
5	2	10	yes	no	no
10	1	5	yes	yes	yes

Table 6.4.: Input parameters, i.e. particle radius r, crack length a and C-rate C, as well the occurrence of the different crack phenomena observed in three dimensional simulations of storage particles during Li insertion.

To reduce computation times the length scale parameter is increased to $l = 0.25\,\mu$m, which allows for a slighty coarser finite element mesh and is still sufficient to resolve the initial cracks. The residual stiffness parameter is kept at $k = 10^{-5}$ and the mesh adaptation scheme again generates a suitable mesh element size close to the crack faces, which is $h \leq (2/5)\,l$. Despite the stronger smoothing of the crack, the refined meshes reach up to 2 million in terms of number of elements and over 199 million in terms of degrees of freedom. The simulations run on either 96 or 128 processors and take up to a maximum of 26 days.

6.2.1 Safe Conditions without Crack Growth

As before, we commence with the simplest situation, i.e. without any crack growth. The corresponding parameters are **r=5 μm, a=1 μm** and **C=5**. The shape of the initial crack is shown in FIG. 6.19.

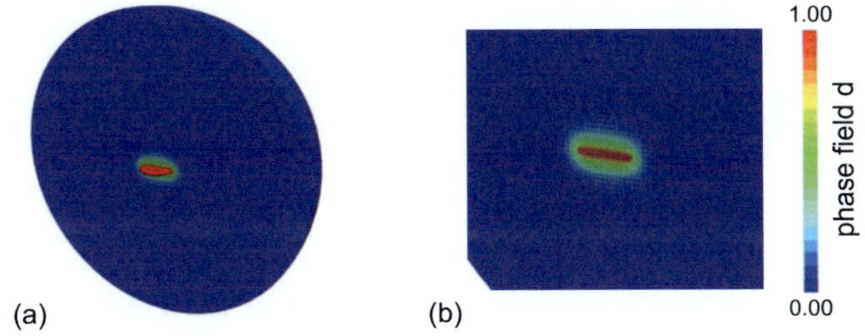

Figure 6.19.: (Left) Initial phase field d representing a precrack of circular shape in a storage particle with $r = 5\,\mu$m and $a = 1\,\mu$m. The volume protruding from the half sphere includes the region having $d > 0.95$. (Right) Magnification around precrack showing the adaptively refined mesh.

The behavior of the characteristic parameters is similar to that described in section 6.1.1 for a two dimensional particle. The state of charge first follows a linear trend until the transition to galvanostatic insertion is reached, as plotted in FIG. 6.20, to be compared to FIG. 6.5. At this point the maximum concentration difference is largest and the characteristic stress parameters peak before beginning to decrease.

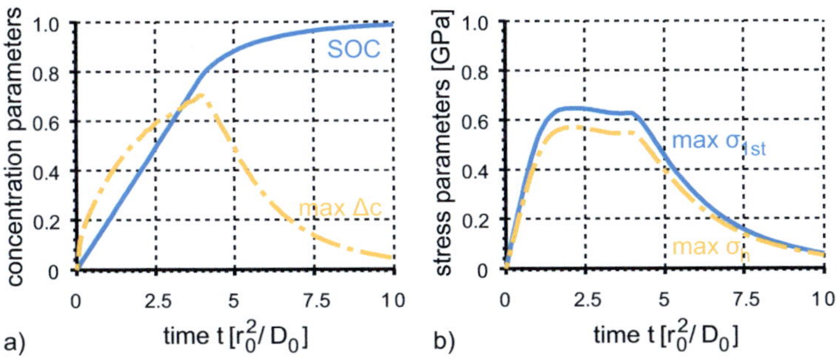

Figure 6.20.: Characteristic results for the 3D insertion example with $r = 5\,\mu$m, $a = 1\,\mu$m and $C = 1$. (a) SOC and maximum concentration difference $\max \Delta c$ in units of maximum concentration c_{max}. (b) Highest value of maximum principal stress $\max \sigma_{1\mathrm{st}}$ and maximum hydrostatic stress $\max \sigma_h$.

The reason that the maximum principal stress hits a plateau at around $t \approx 1.75\,\tau$ while the concentration difference is still growing is twofold. The first reason stems from the specific location at which the maximum and minimum in concentration occur and was also discussed for the two dimensional setting in section 6.1.1. The minimum concentration in the particle that determines the maximum difference, lies in a volume adjacent to the crack faces, as shown in FIG. 6.21. This volume is comparatively small and, most importantly, more compliant than the rest of the material since the crack faces are allowed to separate. It is therefore not affecting the stress in the rest of the body as strongly as the concentration difference between intact regions, i.e. where the order parameter is close to zero. The concentration difference in the intact parts of the particle remains almost constant in the time range of the existence of the stress plateau. Second, we observe that the order parameter increases slightly around the crack, although crack growth is not initiated. This decreases the effective stress due to the prefactor $(1-d)^2$ of the tensile part of the elastic energy, given in equation (2.28).

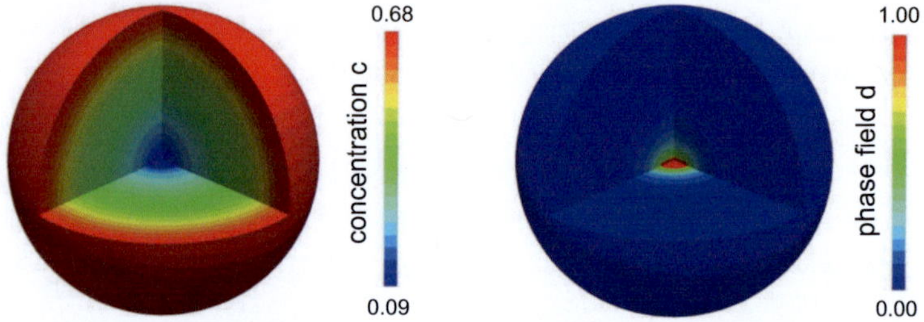

Figure 6.21.: (Left) Concentration c in units of maximum concentration $c_{\max}$ at an intermediate stage of Li insertion in a storage particle with $r = 5\,\mu$m, $a = 1\,\mu$m and $C = 5$. (Right) Order parameter d at the same time step. The comparison shows that the minimum concentration, in dark blue on the left, is located at the faces of the crack, marked by red on the right.

A further difference in FIG. 6.20 when compared to the studies in 2D, specifically the results displayed in FIG. 6.2, is that the highest value of the hydrostatic stress is significantly closer to the highest one for the maximum principal stress. Although both maxima do not have to be located at the same position, the 3D simulation indicates that the maximum hydrostatic stress should occur in a region of high maximum principal stress and that, in contrast to 2D plane strain conditions, the stress state is strongly triaxial with the second and third principal stress being almost equal to the first one.

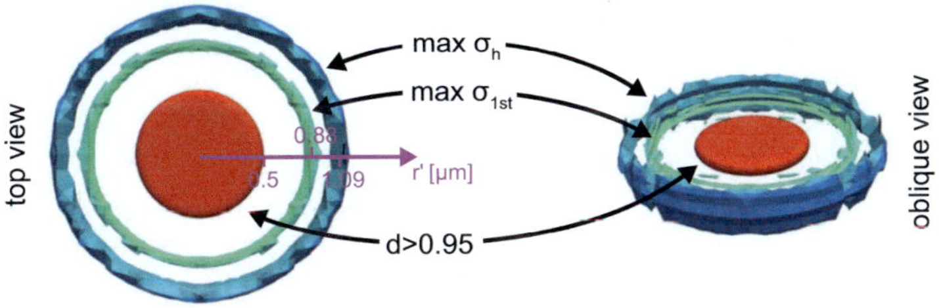

Figure 6.22.: Regions of highest hydrostatic and maximum principal stress, max σ_h and max $\sigma_{1\text{st}}$ respectively. The ring-shaped regions form an extension of the circular crack front, which is defined here by the order parameter $d = 0.95$.

This assertion is confirmed by the fact that the maxima for both quantities are located in ring-like regions around the disk-like initial crack, with the ring of peak maximum principal stress being slightly closer to the crack tip than that for the peak hydrostatic stress, as shown in FIG. 6.22. In the interior of the storage particle close to the crack, the first and second principal stresses are components oriented in the circumferential direction and the third principal stress is orientated in the radial direction. Each principal stress is tensile, as depicted in FIG. 6.23. Close to the faces of the crack, the first principal stress is aligned parallel to them while it acts perpendicular to them close to the ring-like crack tip front. The magnitudes of the principal stresses lie very close to each other (differing below around 8%), which explains the large hydrostatic stress. Closer to the perimeter of the storage particle, the orientation changes. The second and first principal stresses are oriented in the circumferential direction, while the first principal stress acts radially. In this location all stress components are compressive, although the magnitude of the first principal stress is significantly smaller at the boundary, $|\sigma_{1\text{st}}|/|\sigma_{2\text{nd}}| = |\sigma_{1\text{st}}|/|\sigma_{3\text{rd}}| \lesssim 10\%$.

Hence, a small crack in the center of a spherical storage particle merely affects the stress distribution within its center. In the interior the stress state is almost purely hydrostatic and tensile. In contrast, at the perimeter stress is compressive and the radial stress magnitude is significantly lower than that of the hoop stress.

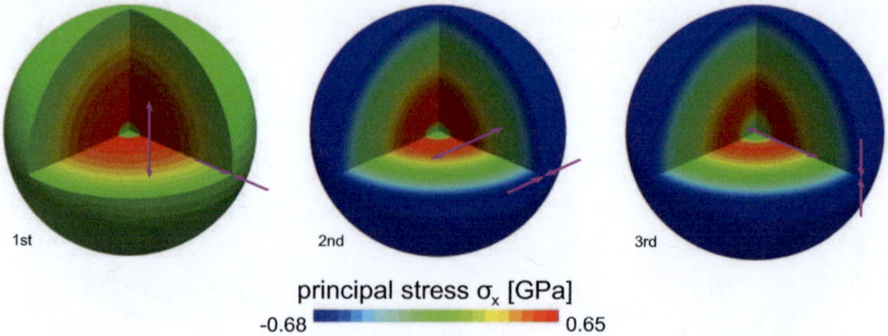

Figure 6.23.: Magnitude and orientation (pink arrows) of principal stresses, taken at an intermediate stage of the Li insertion process, in a storage particle with r=5 μm, a=1 μm and C=5.

6.2.2 Unstable Crack Growth

In order to achieve the critical stress required to initiate crack growth in the storage particle, we increase both the C-rate and the crack diameter and set the parameters to **r=5 μm**, **a=2 μm** and **C=10**. The corresponding crack geometry and Li concentration distribution during Li insertion is shown in FIG. 6.24.

The crack first grows in an unstable fashion and then stops close to the surface. Rotational symmetry is preserved in the results even though an asymmetric perturbation, e.g. a slightly longer extension in one direction than in another, is permitted by the fact that we simulated half of the particle. Due to arrest of the crack in the interior of the particle, the behavior of crack growth and Li diffusion resembles that described in section 6.1.2. The notable similarities are as follows.

First, the Li concentration distribution does not adapt to the modified stress field during the initial phase of crack growth due to the short period of time involved. However, after crack growth has stopped, Li is attracted by the crack tip due to high tensile stress and begins to accumulate around it. This leads to a spot of higher concentration at the crack tip and relative depletion at the surface close to it in 2D, as depicted in FIG. 6.10. In 3D, the crack tip is a ring-shaped front, as illustrated in FIG. 6.26. Hence, the aggregation of Li forms a ring and the lowered concentration on the surface is a circular band, as observable in the top right of FIG. 6.10.

Second, the Li insertion following the unstable crack propagation results in a similar behavior of the highest value of the maximum principal stress, as can be seen by comparison of FIG. 6.7 and FIG. 6.25. The highest value of the maximum principal stress $\max \sigma_{1st}$ first peaks at the onset of unstable growth, then drops rapidly and afterwards grows again to reach a second maximum. In 2D, this process is related to a second period of stable crack growth, as shown in FIG. 6.9. A stable regime is also present in the three dimensional case. However, the distance traveled by the crack tip during stable crack extension ($\Delta a = 0.02\,\mu\mathrm{m}$) is very small in this case.

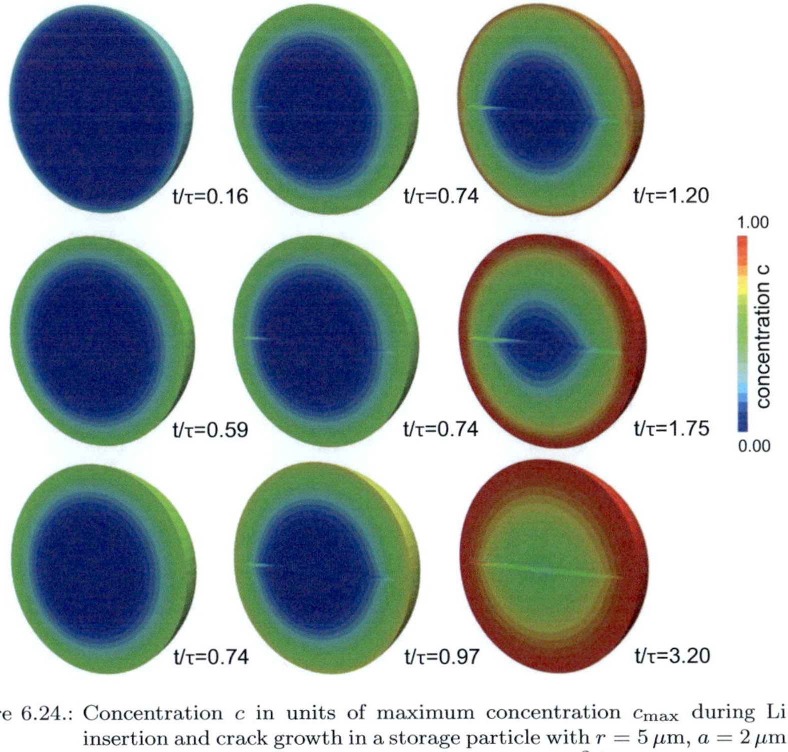

Figure 6.24.: Concentration c in units of maximum concentration c_{max} during Li insertion and crack growth in a storage particle with $r = 5\,\mu\mathrm{m}$, $a = 2\,\mu\mathrm{m}$ and $C = 10$. Times are given in units of $\tau = r_0^2/D_0$. Regions with $d > 0.95$ are removed to show the location of the crack. The opening of the crack is exaggerated by a linear scaling of the nodal displacements.

Remarkably, the second principal stress becomes as large as the first one during the second period of growing stress, as depicted by the blue and dashed red curve in FIG. 6.25. The curves become congruent for $t \gtrsim \tau$, with $\tau = r^2/D_0$. We therefore expect the highest values of first and second principal stress to occur at the same location. The material at the crack is more compliant perpendicular to the crack faces than parallel to them. It is thus unlikely that stress perpendicular to the crack faces and parallel to them acquire the same magnitude at the crack tip. Hence, we expect that the highest tensile stress no longer occurs at the crack tip. In order to confirm this hypothesis we analyze the stress distribution within the storage particle in the following.

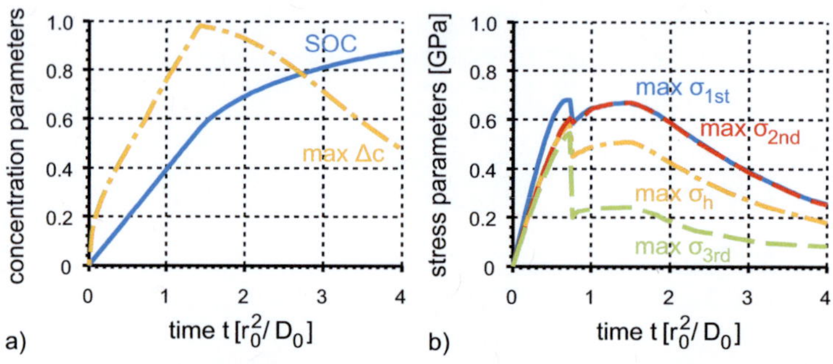

Figure 6.25.: Characteristic results for the 3D insertion example with r=5 μm, a=2 μm and C=10. (a) SOC and maximum concentration difference max Δc in units of maximum concentration $c_{\max}$. (b) Maximum principal stresses max σ_x, $x = \{1\text{st}, 2\text{nd}, 3\text{rd}\}$ and maximum hydrostatic stress max σ_h.

As mentioned above, the crack grows in the form of a disk until it reaches its penultimate radius. This growth relaxes the tensile stress in the plane of the crack perpendicular to its faces. However, as more Li is inserted after unstable growth, the circumferential stress in the material above and below the crack faces further increases. It becomes maximal at a distance of around 2 μm from the particle center along the vertical axis and results in a local maximum of $d \approx 0.26$, as observable in FIG. 6.26. Being located in the axis of symmetry, the stress state is also symmetric and the first and second principal stresses coincide. This explains the congruent curves in FIG. 6.25 and is depicted by the arrows in FIG. 6.26.

Crack growth in a 3D sphere shows similar effects as those obtained under 2D plane strain. These are unstable crack growth, Li attraction at the crack tip and subsequent stable crack growth, given that the particle does not break during unstable growth. However, in contrast to the two-dimensional geometry, even with an extended disk-like crack in the interior, the inhomogeneous volume expansion is so constrained that large magnitudes of tensile stress may occur at locations away from the crack.

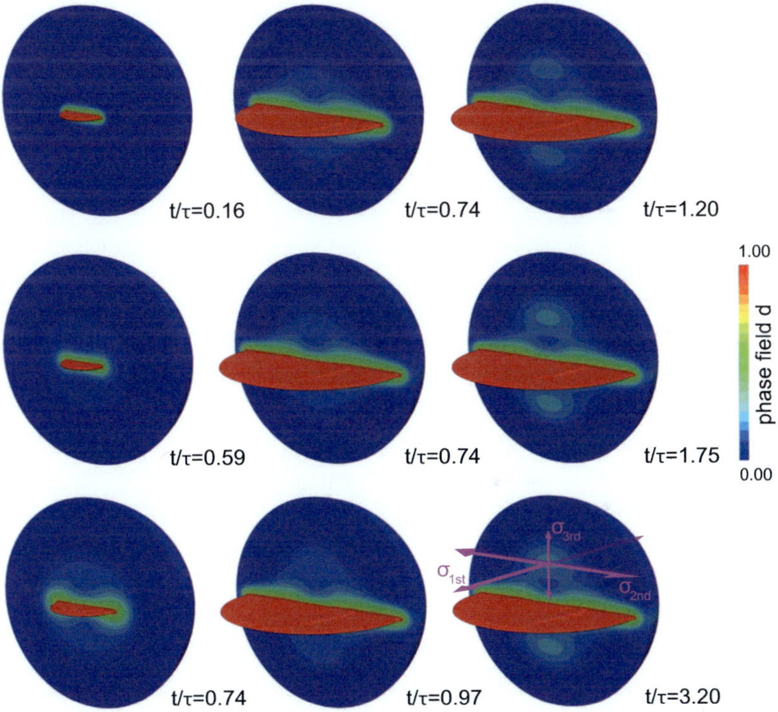

Figure 6.26.: Order parameter d during Li insertion and crack growth in a storage particle with $r = 5\,\mu\text{m}$, $a = 2\,\mu\text{m}$ and $C = 10$. Times are given in units of $\tau = r_0^2/D_0$ and correspond to those shown in FIG. 6.24. The volume protruding from the half sphere includes the region of $d > 0.95$. The pink arrows denote principal stress directions and qualitative magnitudes.

6.2.3 Multiple Initial Cracks

In order to examine whether the large mechanical stress described above, appearing in the vertical symmetry axis at some distance from the crack, is sufficient to initiate crack growth at that location, we introduce a second and third crack above and below the center crack at this location with diameters a in one case and $a/2$ in another. The plane of these cracks is orthogonal to that of the crack at the center of the particle, as can be seen in FIG. 6.27. From the illustrations in the lower row of FIG. 6.27 we observe that the stress generated in the particle after the crack at the particle center has stopped growing, is not sufficient to initiate crack growth of the smaller additional cracks (first column). However, if all cracks are of the same size, the ones closer to the particle surface begin to extend first. Thereafter the stress at the central crack is too small to drive crack growth (second column). If only the cracks above and below the particle center are introduced and the crack at the particle center is absent, both cracks grow and reach a final state that is almost identical to the final state of the central crack when it is present alone (third and fourth column). This means, that although the rotational symmetry around the axis perpendicular to the crack faces is disturbed by the initial cracks, the growth is such that symmetry is restored again.

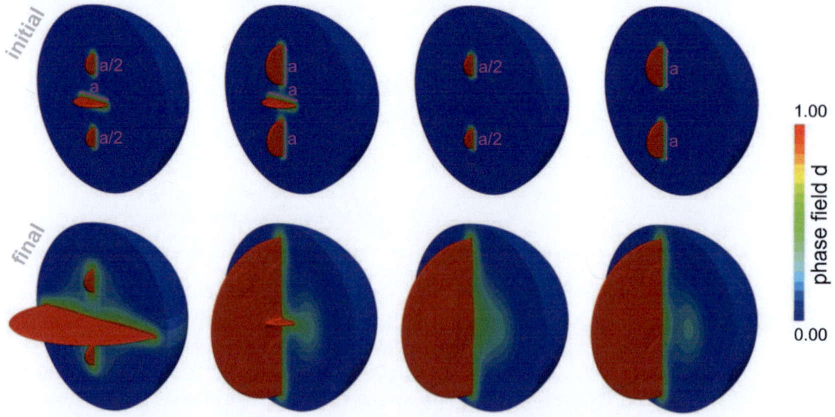

Figure 6.27.: Order parameter d at the initial and final stages of lithiation for different arrangements of cracks in a storage particle with $r = 5\,\mu$m, $a = 2\,\mu$m and $C = 10$. The volume protruding from the half sphere includes the region of $d > 0.95$.

Fracture behavior becomes increasingly complicated with a larger number of initial cracks. However, we find that different locations of the initial crack along the symmetry axes may result in the same final crack geometry due to the approximately spherically symmetric stress state in the particle.

6.2.4 Crack Branching and Particle Breakage

In order to observe crack branching in the 3D particle, the crack tip velocity v_c has to reach high values. In section 6.1, we found that a crack length a close to the critical one a_c and large particle radii, lead to high values of v_c. In 3D we therefore study the parameter combination **r=10 μm, a=1 μm** and **C=5**. Under these conditions, the crack not only branches but breaks the particle into pieces. Observation of the fragments in FIG. 6.28 seems to indicate two cone-like pieces from the top and bottom part of the particle and one ring-shaped piece. However, examining the crack topology more closely, we find that the result becomes more complex.

As depicted in FIG. 6.29, the crack tip at first propagates as a circle of increasing radius. The crack front then branches off into two rings with trajectories forming cone cracks having an angle between them of around 80°. The two rings continue to move until the cone cracks eventually reach the particle surface. However, further branching occurs to relieve the circumferential stress perpendicular to the conical crack faces. In particular, the hoop stress parallel to the original crack front at its location prior to branching remains high and even increases locally during growth of the cone cracks. This hoop stress becomes largest between the upper and lower conical crack faces at a radial distance of around 3.5 μm from the location where the initial branching occurred. The resulting large stress leads to the creation of new splitting cracks perpendicular to the existing ones, as shown in the right column of FIG. 6.29.

The splitting cracks separate the outer ring-like piece into 12 similar sized segments. The angles between these segments lie between 28° and 34°. The reason for the variation in the angle and for the different starting times of the new cracks, observable in the top right of FIG. 6.29, cannot be explained by physical arguments, but is attributed to numerical error. Besides that, the difference in how much each splitting crack has grown in FIG. 6.29 is partially an imaging problem due to the sharp criterion of $d > 0.95$ used in the post-processed plots to specify where the crack surfaces are located. A lower threshold leads to a more uniform appearance for the splitting cracks in terms of size, shape and location.

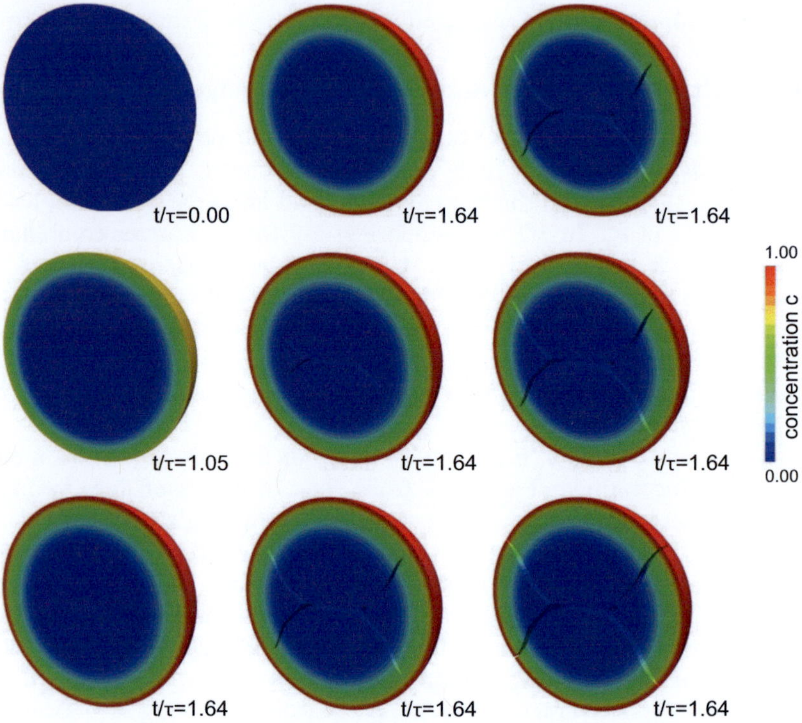

Figure 6.28.: Concentration c in units of maximum concentration c_{max} during Li insertion and crack growth in a storage particle with $r = 10\,\mu\mathrm{m}$, $a = 1\,\mu\mathrm{m}$ and $C = 5$. Times are given in units of $\tau = r_0^2/D_0$. Regions with $d > 0.95$ are removed to show the location of the crack. The opening of the crack is exaggerated by a linear scaling of the nodal displacements.

At an intermediate stage of growth of the splitting cracks (see center right picture in FIG. 6.29) their extensions are all similar. Only the cracks marked with the number 2 are a little less developed. This may result from the larger angle between the faces of those numbered 3, leading to larger stresses for and the earlier onset of the cracks numbered 1. The differences among the crack extensions becomes more pronounced thereafter, as illustrated in the bottom right of FIG. 6.29. The slightly larger sizes of the cracks with numbers 1 and 3 lead to a lower stress at the crack front of that numbered 2 and a consequently slower crack growth. This is to some extent comparable to the findings in [181], where alternating crack lengths of initially equally sized cracks were deduced in the case of thermally induced stress.

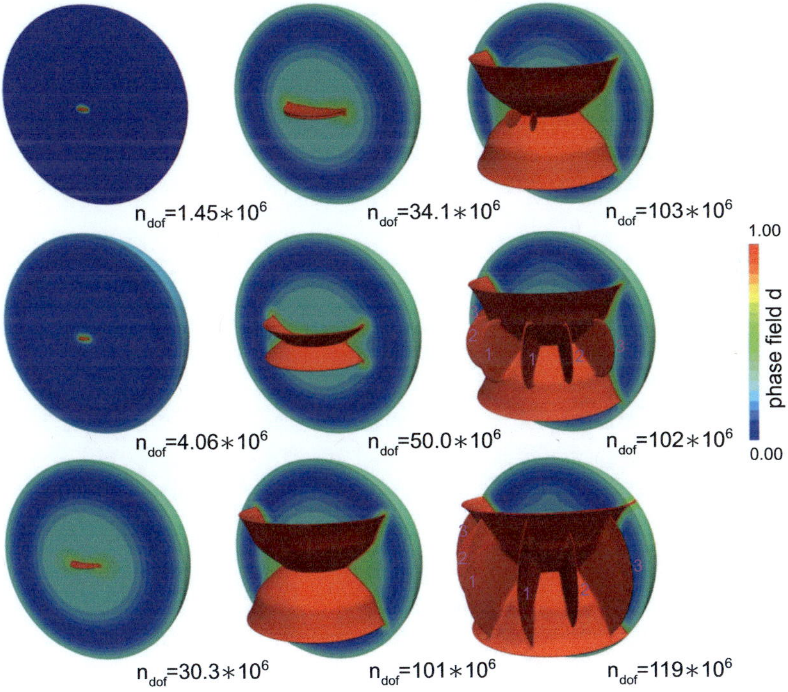

Figure 6.29.: Order parameter d during Li insertion and crack growth in a storage particle with $r = 10\,\mu$m, $a = 1\,\mu$m and $C = 5$. Times correspond to those shown in FIG. 6.28. The volume protruding from the half sphere includes the region of $d > 0.95$. To demonstrate the computational effort required to perform these simulations, the number of degrees of freedoms n_{dof} is given instead of the respective time.

Because of the varying sizes only the cracks with number 1 and 3 reach the particle surface and do so only close to the equator of the particle, as illustrated in FIG. 6.30. Close to the cone crack faces, the vertical cracks do not entirely penetrate to the perimeter. Hence, the circumferential segments are still joined to each other by thin ligaments. Considering the small width of the ligaments, we expect that a subsequent extraction half cycle would tear the segments apart even for low C-rates. In contrast, the parts divided by the faces with number 2 may withstand stressing in subsequent cycles. We cannot prove these speculations due to the computational time required to continue the simulation into a second half cycle. The final number of stable particle fragments is therefore not determined, but may lie between 10 and 14.

We find that in three dimensional particles complex crack patterns can result, when growth of one or several cracks does not relieve enough tensile stress to keep further cracks from growing and when inertia driven effects lead to crack branching. Under these circumstances a storage particle may break into several parts in just one cycle.

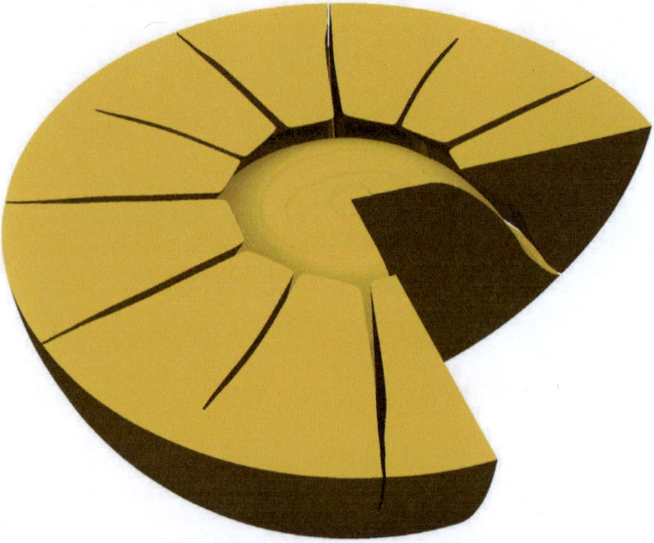

Figure 6.30.: Final crack pattern in a storage particle with $r = 10\,\mu\text{m}$, $a = 1\,\mu\text{m}$ after crack growth during Li insertion with $C = 5$. Regions with $d > 0.95$ are removed to show the location of the crack. The opening of the crack is exaggerated by a linear scaling of the nodal displacements. For better visibility, only a section of the spherical particle is shown.

7 Crack Growth during Lithium Extraction

In the previous chapter, we found that dynamic effects resulting from the inertia of the material have a strong impact on the resulting crack patterns, and may even lead to complete breakage of a storage particle in one half cycle of Li insertion. Here, we focus on the opposite charging condition and examine the dependence of unstable crack growth on particle radius and initial crack length for different rates of Li extraction.

For the sake of comparability with our previous investigation and in order to reduce the complexity of the study, we vary the same set of parameters as in the study on crack growth during Li insertion, presented in chapter 6, and focus on spherical (3D) or cylindrical (2D plane strain) particles of outer radius r, as illustrated in FIG. 7.1. The boundary condition for the Li diffusion problem is applied in a mode that is analogous to constant current constant voltage (CCCV) operation of a battery cell. First, a constant flux $\bar{J}_0$ is imposed on the perimeter until the minimum concentration c_{min} is reached. Throughout this work, we set $c_{\mathrm{min}} = 0$. Thereafter the concentration c is held fixed. The two consecutive steps are also known as galvanostatic and potentiostatic discharge respectively. The magnitude of the imposed flux $\bar{J}_0$ is prescribed in the form of the C-rate C, i.e. the inverse time in hours theoretically required to completely charge or discharge the particle. With V and S denoting the volume and surface of the storage particle respectively, it can be expressed through

$$|\bar{J}_0| = \frac{V}{S} c_{\mathrm{max}} C/[3600\,\mathrm{s}]. \tag{7.1}$$

For Li extraction, the largest tensile stress occurs at the surface of a particle, as shown, for example, in [129]. This stress might lead to fracture starting from voids or pores close to the surface and incisions or defects on the surface. Initial cracks are therefore introduced at the particle surface with the shape of a line (2D) or a half-disk (3D) with length or radius a. As in the study on Li insertion, material parameters are chosen by using values for $LiMn_2O_4$, as given in TAB. 6.1 of chapter 6.

Following the findings in the assessment of the phase field method for crack growth, presented in chapter 4, the initial cracks are introduced as heterogeneities in the phase field through an initially non-zero history field $\mathcal{H}_0$, as given by

$$\mathcal{H}_0(x,y) = \alpha\, e^{-(y/\beta)^2} \times \begin{cases} 1, & s < 0 \\ e^{-(s/\beta)^2}, & s \geq 0, \end{cases} \tag{7.2}$$

where $\alpha = 10^4$ and $\beta = l/10$. The local distance to the crack tip is defined by $s = |x| - (r - a)$ in two dimensions and $s = \sqrt{(|x| - r)^2 + y^2} - a$ in three dimensions. The so defined initial history field introduces two cracks on opposite sides of the storage particle and therefore preserves the mirror symmetry with respect to the horizontal and vertical plane through the center of the storage particle, as shown in FIG. 7.1.

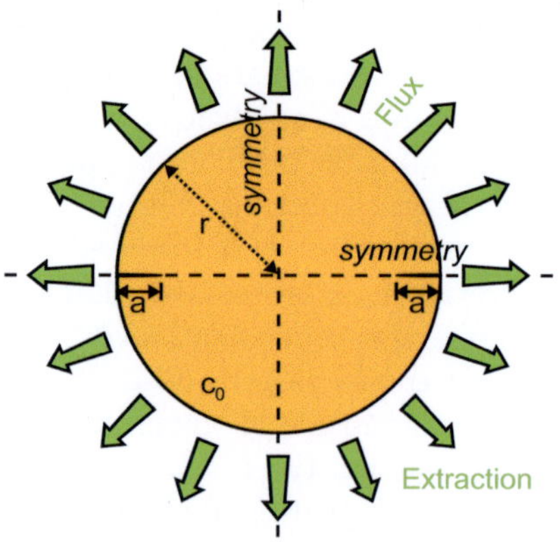

Figure 7.1.: Geometry and boundary conditions for the Li deintercalation study. The geometry is either 2D plane strain in section 7.1 or, under neglect of the initial cracks, 3D axisymmetric in section 7.2. In the three dimensional geometry, the initial cracks have the shape of half-disks. This is illustrated, for example, in FIG. 7.15.

This allows savings in computation time by simulation of only one quarter of the particle in two dimensions, and one half of it in three dimensions. The simulations are started with maximum concentration, i.e. $c_0 = c_{\max}$, and stopped when the state of charge (SOC), describing the mean concentration in units of maximum concentration $c_{\max}$, reaches a level below 0.01%.

In the investigations on Li insertion, initial crack growth always occurs in a markedly unstable way. Only in the case of crack growth termination inside the storage particle, is stable crack propagation observed after further Li insertion. As will be shown in the following sections, the regime of unstable crack growth is generally less pronounced when the crack starts to extend from the surface towards the center during Li extraction. For this reason, the transition from unstable to stable growth is less abrupt, but instead rather smooth. Because of the smoother transition, the distinction between stable and unstable crack growth is less obvious and the method of using the sudden and dramatic decrease of the time step to differentiate between unstable and stable crack propagation, used in chapter 6, cannot always be used in the case of Li extraction.

Instead of using the drop of the time step to identify when unstable crack growth commences, we find that the time rate of change of the total crack surface area $\dot{\mathcal{A}}_{\text{crack}}$ is a suitable alternative indicator for differentiating between stable and unstable crack growth. As described in the context of equation (2.23), the phase field method approximates the crack surface by integration over the crack surface density through

$$\mathcal{A}_{\text{crack}} \approx \int_{\mathcal{B}} \gamma_l(d) \mathrm{d}V, \tag{7.3}$$

where the crack surface density is defined in equation (2.23) by

$$\gamma_l(d) = \frac{1}{2l} d^2 + \frac{l}{2} |\nabla d|^2. \tag{7.4}$$

In the case of the 2D plane strain geometry, the total crack surface area is proportional to the length of each crack via $\mathcal{A}_{\text{crack}} = 2ah$, where h is an arbitrary thickness associated with the plane strain condition. We set it to $h = 1\,\mu\text{m}$. For one single crack or multiple cracks of equal length, evaluation of the time rate of change of the total crack surface area $\dot{\mathcal{A}}_{\text{crack}}$ is therefore equivalent to observation of the crack tip velocity. However, evaluation of the crack tip velocity requires cumbersome tracking of the crack tip in an additional post-processing step, and is therefore avoided through use of the time rate of change of the total crack surface area.

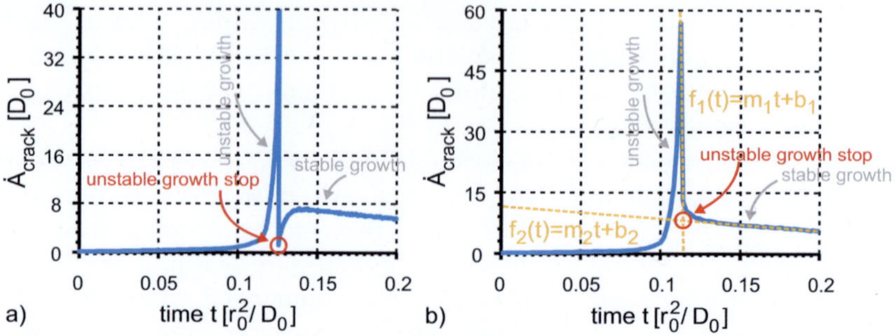

Figure 7.2.: Time rate of change of the total crack surface area $\dot{A}_{\text{crack}}$ during Li extraction in a storage particle with radius $r = 5\,\mu\text{m}$, initial crack length $a = 0.1\,\mu\text{m}$ and C-rates of (a) $C = 12$ and (b) $C = 13$. For better visibility of the transition from unstable to stable crack growth, the vertical axis in (a) is adjusted such that it does not cover the full range of the time rate of change of the total crack surface area. Linear fits in the stable und unstable regime are shown in orange in (b), and their use is explained in the text. The time is given in units of r_0^2/D_0 with $r_0 = 1\,\mu\text{m}$.

The suitability of $\dot{A}_{\text{crack}}$ as a measure to differentiate between stable and unstable crack growth regimes shall be explained by means of FIG. 7.2, which shows the time history of $\dot{A}_{\text{crack}}$ during Li extraction for two different C-rates at equal radius and crack length. In the case of unstable crack growth, the extension of a crack does not release sufficient elastic tensile energy to stop further growth. The time rate of crack extension is therefore only controlled by inertia effects. The time scale of these effects is in the range of, for example, the time that a sound wave requires to travel from one side of the storage particle to the other side. In contrast, during stable crack growth, the crack length is always in equilibrium with the applied load in the sense that the energy release rate is equal to its critical value, i.e. $G = G_c$. Hence, in order to drive stable crack growth in a continuous way, the external load has to be increased steadily. In terms of the Li concentration, this means that further Li has to be extracted or inserted. In the case of stable crack growth, the time rate of change of the total crack surface area $\dot{A}_{\text{crack}}$ is therefore closely linked to the charge or discharge rate. Consequently, the time scale corresponding to stable crack growth is significantly longer than that associated with unstable crack growth. Vice versa, the time rate of change of the total crack surface area $\dot{A}_{\text{crack}}$ is substantially larger during unstable crack growth. This behavior is reflected by the sharp increase of $\dot{\mathcal{A}}_{\text{crack}}$ at around $t \approx 1.2\,\tau$ in FIG. 7.2 a) and $t \approx 1.1\,\tau$ in FIG. 7.2 b), and a notably smaller magnitude of $\dot{\mathcal{A}}_{\text{crack}}$ for $t \gtrsim 1.25\,\tau$ and $t \gtrsim 1.15\,\tau$ respectively. As in chapter 6, the characteristic time scale is defined as $\tau = r_0^2/D_0$ with $r_0 = 1\,\mu\text{m}$. The intermediate point of almost vanishing rate of the total crack surface, marked by the kink in FIG. 7.2 a), can be understood by the findings in the study on crack growth during Li insertion, for example, described in section 6.1.2. During crack growth, the material adjacent to the crack faces is accelerated away from each other due to the increasing crack opening and linear momentum builds up. The deceleration of the material then leads to an inertia induced stress field in addition to the stress from the volume expansion, when crack growth is slowed down. This stress field drives the crack to grow farther than its equilibrium length during stable growth. The energy release rate then falls below its critical value and the crack stops to grow. This is marked by the red circle in FIG. 7.2 a). Only after further Li is extracted from the storage particle, the mechanical stress at the crack tip becomes again large enough to drive further crack growth, now in a stable way. The time of $\dot{\mathcal{A}}_{\text{crack}} \approx 0$ thus marks the transition from stable to unstable growth. We use it to determine the distance of stable and unstable crack propagation.

The variation in the shapes of the graphs in FIG. 7.2 a) and b) results from a less pronounced unstable crack growth regime in the latter case. When the extent of unstable crack growth is short, the maximum crack tip velocity is comparatively small, i.e. $\dot{\mathcal{A}}^{a)}_{\text{crack}} > \dot{\mathcal{A}}^{b)}_{\text{crack}}$. As a result, the momentum build-up in the material adjacent to the crack faces is negligible and the additional crack extension driven by inertia forces is marginal. In this case, the small extra distance leading to a preliminary termination of crack growth for which $\dot{\mathcal{A}}_{\text{crack}}$ drops to zero is not resolved within the simulations for the chosen numerical parameters. In order to identify the transition from unstable to stable growth nonetheless present in such a case, we use the intersection of the linear fits $f_1(t)$ and $f_2(t)$, as illustrated in FIG. 7.2 b). For shorter extents of total unstable crack growth, the graph becomes progressively smoother, so that the distinction between the initial peak, representing the onset of unstable growth, and the subsequent stable part of the graph, is less clear. We therefore use the slopes m_1 and m_2 in a criterion to determine if crack growth is exclusively stable ($m_1/m_2 < 2$) or if an unstable crack growth regime exists ($m_1/m_2 \geq 2$). The criterion provides a consistent

operational measure for deciding whether unstable crack growth occurs. However, we note that the choice for the threshold of the ratio m_1/m_2 is somewhat arbitrary and the determination of the precise transition between unstable and stable growth is less accurate in the case of smaller extents of unstable growth.
Before going into the details of the study, we comment on the flux condition at the crack surfaces. As described in the formulation of the model, presented in section 2.3, the diffusion of Li is only influenced by the order parameter d through modifications to the stress field. Hence, introducing the crack as a heterogeneity in d does not specify the crack surfaces in the sense that Li can be inserted through them. This may not describe the actual physical situation perfectly, but nevertheless provides a good first approximation in absence of physically sound information on the precise Li transport conditions inside the crack; any assumption that we might make in this regard is currently not supported by experimental observations. However, we emphasize that our simplified assumption in the case of extraction differs in its implications from when we used it in our study of Li insertion in chapter 6. In that case a more elaborate condition for diffusion at the crack surface could be neglected on physical grounds due to the symmetries in the problem, the short time scale of unstable crack growth and the initial location of the crack in the interior of the storage particle. While the argument of symmetry remains valid in the following study, the crack now lies at the perimeter of the storage particle and crack growth towards the center of the storage particle is comparatively slow. As a consequence, electrolyte may flow into the crack and fill the void volume. Thus, although Li diffusion across the crack can still be neglected for reasons of symmetry, Li-ions could, in principal, deintercalate at the crack faces and leave the crack through transport in the electrolyte.

Parameters			crack behavior		
r [μm]	a [μm]	C	growth	unstable	add. cracks
2	0.1...1.0	1...10	no	no	no
3	0.1...1.0	1...5	no	no	no
3	0.1	10	yes	yes	no
3	0.5	10	yes	no	no
3	1.0	10	yes	no	no
5	0.1...1.0	1	no	no	no
5	0.1	5	yes	yes	no
5	0.5	5	yes	no	no
5	1.0	5	yes	no	no
5	0.1...10	10	yes	no	no
10	0.1	1	yes	yes	no
10	0.1	5	yes	yes	no
10	0.1	10	yes	no	no
10	0.5	1	yes	yes	no
10	0.5	5	yes	no	no
10	0.5	10	yes	no	no
10	1.0	1...10	yes	no	no
15...20	0.1	1	yes	yes	no
15...20	0.1	5	yes	no	yes
15...20	0.1	10	yes	no	yes
15...20	0.5...1.0	1	yes	no	no
15...20	0.5...1.0	5	yes	no	yes
15...20	0.5...1.0	10	yes	no	yes

Table 7.1.: Input parameters, i.e. particle radius r, crack length a and C-rate C, as well as type of crack growth, i.e. entirely stable or with an initial unstable phase, and occurrence of additional cracks in the 2D extraction study. Dots between two numbers imply that the corresponding line is valid for the entire range in between, e.g. storage particles with $r = 2\,\mu$m and initial crack length a between 0.1 and 1.0 μm show no crack growth when Li is extracted with a C-rate between 1 and 10.

7.1 Cylindrical Particles in 2D under Plane Strain

The particle size is varied between $r = 2\,\mu\text{m}$ and $20\,\mu\text{m}$, the crack sizes are $a = 0.1\,\mu\text{m}$, $0.5\,\mu\text{m}$ and $1\,\mu\text{m}$ and C-rates range between $C = 1$, 5 and 10, as shown in TAB. 7.1. The length scale parameter is typically set to $l = 0.05\,\mu\text{m}$, the residual stiffness parameter is chosen as $k = 10^{-5}$ and the mesh adaptation scheme is configured such that the mesh element size close to the crack faces is $h \leq (2/5)\,l$. To obtain even more accurate results in the study of unstable crack growth with varying initial crack lengths and C-rates, the length scale parameter is decreased to $l = 0.01\,\mu\text{m}$ and $l = 0.025\,\mu\text{m}$ respectively and the residual stiffness is set to $k = 10^{-7}$ and $k = 10^{-5}$ respectively.

Depending on the particle radius, the number of initial global mesh refinements and the stage of crack growth, the number of elements ranges from 843 to 53 532, or in terms of degrees of freedom it ranges from 17 650 to 1 085 565. Simulations are typically run on either 16 or 32 processors and last between 30 minutes and around 200 hours, heavily influenced by system sizes and crack behavior.

With respect to TAB. 6.2, given in the study on crack growth during Li insertion of chapter 6, and TAB. 7.1, we find that it generally requires less severe conditions to drive crack growth during Li extraction in comparison to Li insertion. For example, a surface crack with length $a = 0.1\,\mu\text{m}$ in a storage particle with radius $r = 3\,\mu\text{m}$ grows during Li extraction with $C = 10$. In contrast, even in a storage particle with radius $r = 20\,\mu\text{m}$, a crack of length $a = 0.2\,\mu\text{m}$ does not commence growing under Li insertion at the same C-rate. Furthermore, for Li extraction, the existence of crack growth does not seem to be affected by the initial crack length a as strongly as for Li insertion. Indeed, within the chosen set of parameters, the question whether crack growth takes place does not depend on a, but instead only on the particle radius and C-rate. However, the type of initial growth, i.e. whether it is stable or unstable, is influenced by the crack size a. We return to these observations and consider them in more detail after first investigating in the next section the principal reason for stress generation during Li extraction in situations without crack growth.

7.1.1 Safe Conditions without Crack Growth

Safe conditions, in the sense that no crack growth takes place in the storage particle, are found for particle radii of $r = 2\,\mu\text{m}$, $r = 3\,\mu\text{m}$ (except for $C = 10$) and $r = 5\,\mu\text{m}$ (except for $C = 5$ and $C = 10$), as listed in TAB. 7.1. The strong impact of particle radius and C-rate on the occurrence of crack growth is explained by reasons similar to those invoked in the case of Li insertion, described in chapter 6. The characteristic time scale for Li to diffuse into the particle is approximately given by r^2/D_0. Thus, as r and C are increased, the concentration gradient becomes stronger and higher mechanical stress is generated. A characteristic parameter related to the inhomogeneous Li concentration occurring in the storage particle is the maximum concentration difference $\max \Delta c$. It serves as a suitable indicator for the interplay of Li diffusion and mechanical stress and its evolution during Li extraction is discussed at corresponding passages in this work. In the following study, the correlation between the maximum concentration difference and the stress obtained during Li extraction will be explored. To illustrate the role of the initial crack size a on the mechanical stress in the particle, we take a look at Li extraction in two particles of the same size and at the same C-rate, but with different initial crack lengths. For this, we examine the parameter combinations **r=3 μm, a=0.1 μm, C=5** and **r=3 μm, a=1.0 μm, C=5**. Results for specific parameters are plotted in FIG. 7.3.

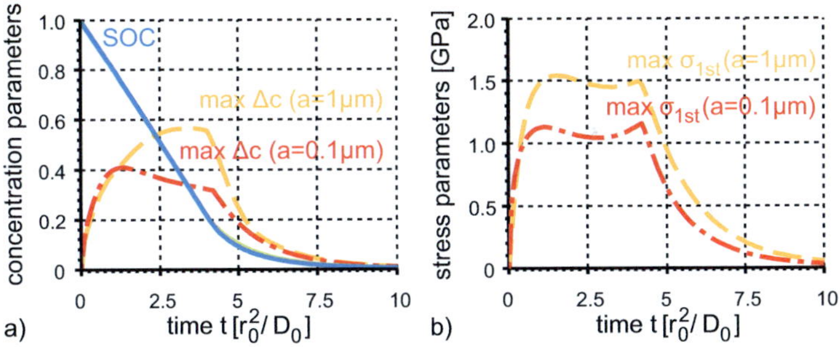

Figure 7.3.: Characteristic results for Li extraction with $r = 3\,\mu\text{m}$, $C = 5$ and $a = 0.1\,\mu\text{m}$, or $a = 1.0\,\mu\text{m}$ as a function of time. (a) State of charge (SOC) and maximum concentration difference $\max \Delta c$ in units of maximum concentration c_{max}. No difference between the graphs for the SOC for $a = 0.1\,\mu\text{m}$ and $a = 1.0\,\mu\text{m}$ is recognizable. (b) The highest value of the maximum principal stress $\max \sigma_{1\text{st}}$ during Li deintercalation.

We find that the highest value of the maximum principal stress $\max \sigma_{1\text{st}}$ present in the particle during extraction of Li is higher by up to 34% in the case of the larger initial crack length. This difference results due to the following reason. During Li extraction, the concentration of Li decreases and the particle shrinks. This shrinkage is most pronounced close to the particle surface, where the largest amount of Li is being extracted. Consequently, tensile hoop stress develops on the surface and compressive radial stress is generated in the interior. For a larger initial crack length, the portion of the particle that lies outside of the radius $r - a$ and generates tensile stress at the crack tip is bigger, as shown in orange color in FIG. 7.4. Thus, more contracting material acts on the crack tip and the resulting tensile stress is increased, as can be seen in FIG. 7.3. However, as a is increased beyond a certain value, the stress at the crack tip becomes smaller again since most of the volume contraction is relaxed due to the larger crack opening and since the volume with higher Li content in the interior becomes smaller.

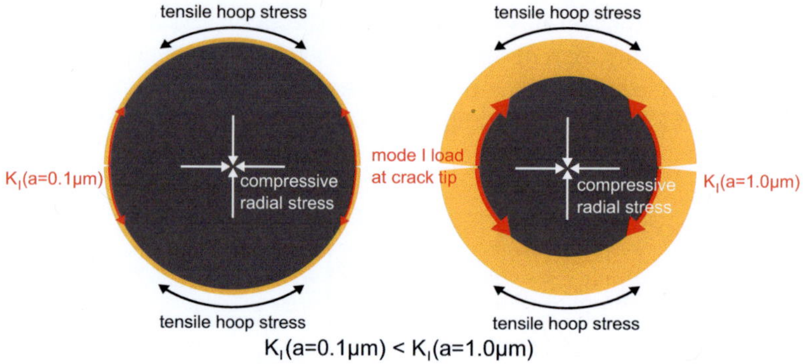

Figure 7.4.: Schematic illustration of the mechanical stress state in the storage particle during Li extraction for two initial crack sizes. The exact stress distribution is complicated by the inhomogeneous Li concentration and the presence of the crack. However, in a coarse approximation, the region shaded gray at the core of the particle is in radial compression while that having the lighter shade (orange) is subject to circumferential tension.

The effect of the lowered stress away from the crack tip on the Li concentration distribution, resulting from the larger crack opening with increased crack length, can be observed in FIG. 7.5. Due to the larger initial crack length, both the tensile stress on the surface and the compressive stress in the interior are reduced. Hence, the gradient in the hydrostatic stress pointing from the surface towards the center is decreased as well. As a consequence, the mechanical driving force for Li flux, as given in equation

(2.20), is less pronounced and a larger concentration difference between surface and interior builds up. This can be recognized by the deeper blue at the surface and the slightly more yellow color in the center.

Apart from this global effect, the higher tensile stress at the crack tip leads to a significant increase in the local concentration at the crack tip, which becomes even larger than in the interior. This strongly localized accumulation of Li may trigger undesired phase transitions in the material, eventually leading to voltage or capacity fade. This effect was already found in the case of Li insertion.

The presence of a surface crack alters the Li concentration distribution during extraction in two ways. First, a global effect on the mean concentration gradient results from the relaxation of tensile stress at the surface due to the crack opening. And second, high tensile stress at the crack tip attracts Li in a strongly localized way.

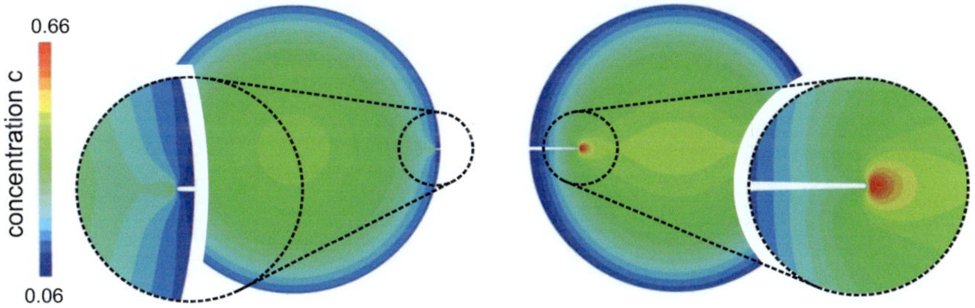

Figure 7.5.: Concentration c in units of maximum concentration $c_{\max}$ during Li extraction for $r = 3\,\mu\text{m}$, $C = 5$ and $a = 0.1\,\mu\text{m}$ (left), as well as $a = 1.0\,\mu\text{m}$ (right). The time selected for the plots is $t = 3.5\,\tau$, with $\tau = r_0^2/D_0$. Areas with $d > 0.95$ are removed to show the location of the crack. The opening of the crack is exaggerated by a linear scaling of the nodal displacements.

7.1.2 Stable Crack Growth

As shown in TAB. 7.1, crack growth only takes place when the combination of radius, crack length and C-rate exceeds a certain threshold. To study the interplay of Li diffusion and mechanical stress during crack growth, we therefore increase the C-rate and investigate the parameter combination $\mathbf{r=3\,\mu m}$, $\mathbf{a=1\,\mu m}$ and $\mathbf{C=10}$. The physical behavior discussed in the following is representative of all input parameters, that neither lead to unstable crack growth nor creation of additional cracks.

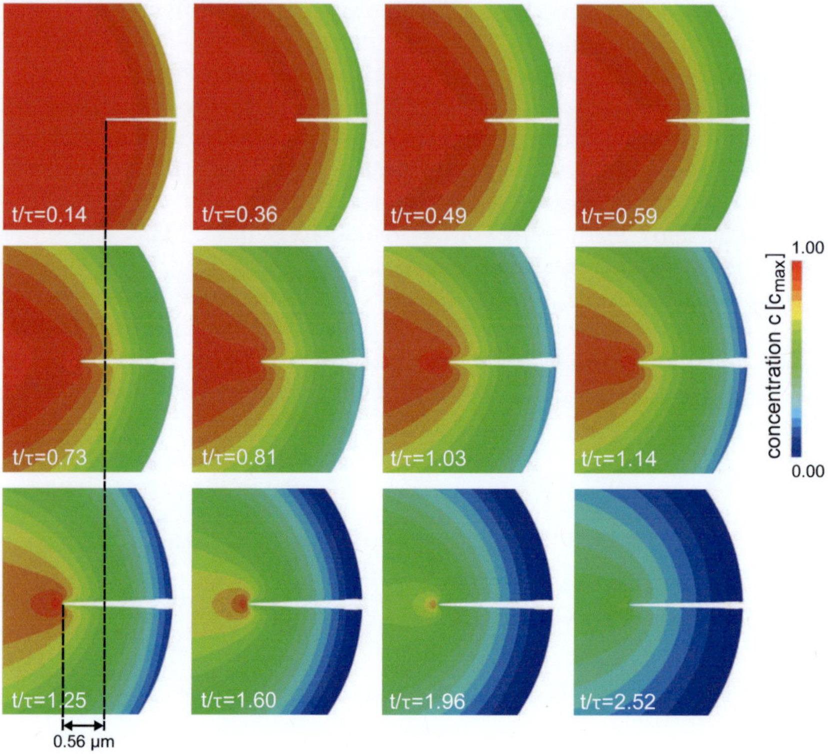

Figure 7.6.: Li concentration in units of maximum concentration c_{max} redistributing around the crack tip during crack growth in a storage particle with $r = 3\,\mu\mathrm{m}$, $a = 1\,\mu\mathrm{m}$ and $C = 10$. Dashed lines are drawn to aid recognition of the extension of the crack during growth. Times are given in units of $\tau = r_0^2/D_0$.

As shown in FIG. 7.7, the highest value of the maximum principal stress rises to a maximum of $\max \sigma_{1\mathrm{st}} \approx 1.65\,\mathrm{GPa}$ at $t \approx 0.45\,\tau$, thereupon crack growth is initiated. It then follows a plateau until the crack stops growing at $t \approx 1.72\,\tau$. Since the transition from galvanostatic to potentiostatic Li extraction occurs at $\tau_{\mathrm{galvano}} \approx 1.48\,\tau$, the stress at the crack tip is still high enough to induce further crack propagation within the time period between $1.48\,\tau$ and $1.72\,\tau$ even though the maximum concentration difference in the particle has begun to decrease.

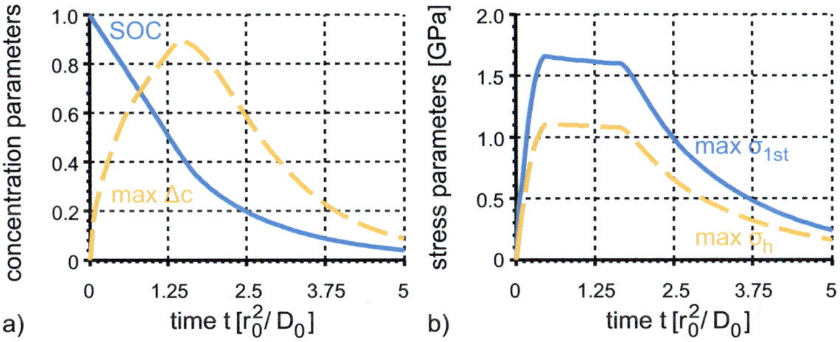

Figure 7.7.: Characteristic results for Li extraction with $r = 3\,\mu$m, $a = 1\,\mu$m and $C = 10$. (a) State of charge (SOC) and maximum concentration difference $\max \Delta c$ in units of maximum concentration c_{max}. (b) The highest value of the maximum principal stress $\max \sigma_{1\text{st}}$ and maximum hydrostatic stress $\max \sigma_h$ during Li deintercalation.

The maximum concentration difference at the transition from galvanostatic to potentiostatic Li extraction is $\max \Delta c \approx 0.9\, c_{\text{max}}$. At this time, the minimum concentration is located at the surface and the maximum concentration is found at the crack tip, which acts as a center of attraction for Li due to the large tensile stress at this location. This effect can also be observed in FIG. 7.6. After the boundary of the region of maximum concentration $c = c_{\text{max}}$, shown in deep red color, has been passed by the crack tip, the concentration dependent prefactor $c\,(c - c_{\text{max}})$ of the mechanical driving force for diffusion, given in equation (2.20), becomes non-zero at the crack tip. As a consequence, the Li concentration adapts to the stress field around the crack tip (cf. plots at $t \approx 0.14\,\tau$ and $t \approx 0.36\,\tau$). A location of high Li concentration forms at the crack tip which moves with the crack tip during its propagation. This is in contrast to the unstable crack growth that occurs for Li insertion, described in chapter 6, where the propagation of the crack tip is typically too fast for Li to follow. In this example, crack propagation and Li extraction thus take place on the same time scale. This is also visible in FIG. 7.7, where the time range of crack growth, marked by the plateau in the highest value of the maximum principal stress, covers a substantial part of the overall extraction time. The chosen combination of parameters therefore presents an example of entirely stable crack growth. This statement is also in agreement with the criterion derived in the introduction of this chapter, which takes into account the slope of the time rate of change of the crack surface area $\dot{A}_{\text{crack}}$. That is, for this combination of parameters we find that $m_1/m_2 < 2$.

Hence, the chosen parameter range indicates that crack growth during Li extraction, starting from a surface crack, occurs under less severe conditions, in the sense of lower particle radii, crack length and C-rates, as compared to growth of a central through crack during Li insertion. Also, crack propagation does not necessarily take place in an unstable way when Li is being extracted.

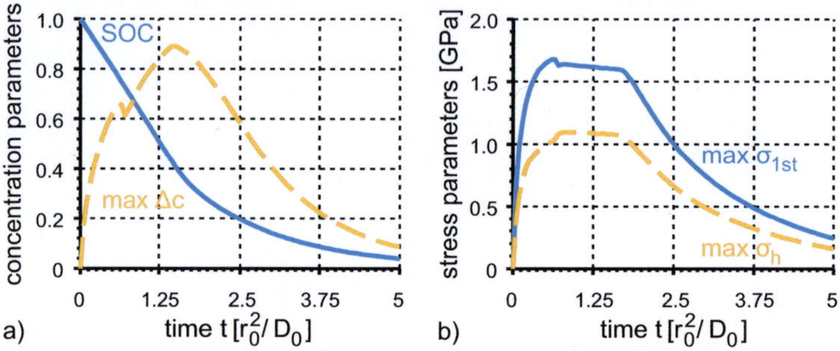

Figure 7.8.: Characteristic results for Li extraction with $r = 3\,\mu$m, $a = 0.1\,\mu$m and $C = 10$. (a) State of charge (SOC) and maximum concentration difference $\max \Delta c$ in units of maximum concentration $c_{\max}$. (b) The highest value of the maximum principal stress $\max \sigma_{1\text{st}}$ and maximum hydrostatic stress $\max \sigma_h$.

7.1.3 Unstable Crack Growth

In the consideration of Li insertion in cases of unstable crack growth, presented in chapter 6, it was found that for small initial cracks, i.e. $a \ll r$, the mechanical load on the crack increases with the crack length a. Hence, if the mechanical load induced by the Li concentration distribution is sufficient for initiation of growth of a crack with length a, it is also sufficient for initiation of growth of a crack with length $a' > a$. For this reason, the extension of the crack does not lead to a new equilibrium in the sense that $G \leq G_c$ and crack growth occurs in an unstable way. This is not observed in the example of stable growth presented in the previous section. During Li insertion, stable crack growth only takes place for large initial crack sizes, where the assumption of $a \ll r$ is not valid anymore. In order to check whether unstable growth of small initial cracks also takes place during Li extraction, we decrease the crack length and consider the parameters **r=3 μm**, **a=0.1 μm** and **C=10**. The corresponding concentration and stress results are plotted in FIG. 7.8.

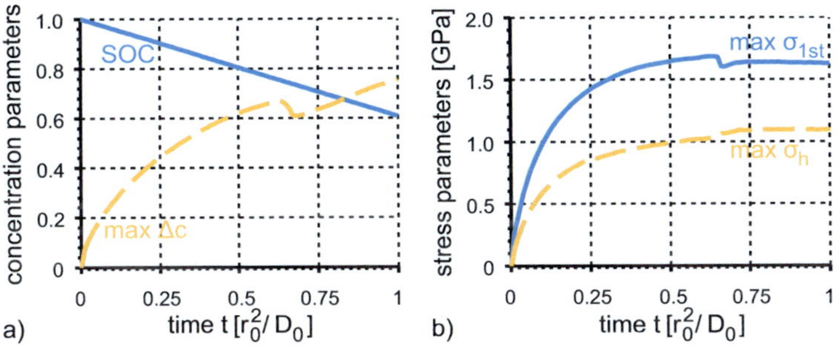

Figure 7.9.: Magnified view of the characteristic results in FIG. 7.8 for early time. (a) State of charge (SOC) and maximum concentration difference max Δc in units of maximum concentration c_{max}. (b) The highest value of the maximum principal stress max $\sigma_{1\text{st}}$ and maximum hydrostatic max σ_h.

The peak and subsequent drop in the highest value of the maximum principal stress around $t \approx 0.65\,\tau$ indicate that a regime of unstable crack growth exists for the chosen combination of parameters. In order to show this detail more clearly, a magnified view of the diagrams is plotted in FIG. 7.9. The drop in the highest value of the maximum stress results from the additional extent of crack growth induced by inertia forces. This effect was discussed in the context of FIG. 7.2 and first found in our study of crack growth during Li insertion. The crack propagates further than it would if only promoted by the mechanical stress resulting from the concentration distribution present at that time. Thus, as inertia forces have relaxed, the stress at the crack tip temporarily falls below the magnitude required to drive further crack growth.

The comparatively small time scale of the unstable crack growth results in a sharp kink in the curve describing the maximum concentration difference, as shown in FIG. 7.8 a) and FIG. 7.9 b). The tensile stress in the outer shell of the particle is smallest at the crack faces, since some of the tensile stress is relaxed due to the crack opening. Consequently, the gradient of the hydrostatic stress points away from the crack faces. As a result, Li is repulsed from these regions and the local Li concentration decreases. This is shown, for example, by the deep blue region of the last contour plot in the upper row of FIG. 7.10. The maximum difference in concentration $\max \Delta c$ thus takes place between the center of the storage particle and the material adjacent to the crack faces. This Li concentration distribution remains almost constant during unstable crack growth due to the high velocity of the process, see middle row in FIG. 7.10. However, after unstable propagation ceases, the crack tip is further away from the surface and the stress in the outermost shell of the particle is relaxed. The gradient in the hydrostatic stress that previously forced Li to move away from the crack faces is decreased, so that Li now distributes itself more homogeneously at the particle surface (cf. plots at $t \approx 0.65\,\tau$ with $t \approx 0.66\,\tau$ and $t \approx 0.68\,\tau$). The Li concentration in the material adjacent to the crack faces thereby increases so that the maximum concentration difference $\max \Delta c$ becomes smaller. Only after more Li is extracted at the surface does the maximum concentration difference rise again.

Due to continued Li extraction after termination of unstable crack growth, the stress again becomes large enough to drive crack growth further. However, crack propagation now takes place in a stable fashion, as shown in the bottom row of FIG. 7.10. The separation into a regime of unstable and stable crack growth can also be observed in FIG. 7.11, where we plot the time history of the crack length a. The small time scale of unstable crack growth is reflected by the step-like shape of the graph of the crack length. Following this increase in crack length, it remains constant for a short period of time before slowly rising to its final value as a result of stable crack growth. This behavior is similar to what was reported by Bahr et al. in [182] in the case of surface cracks in a specimen that undergoes thermal shock. Depending on the temperature difference and initial crack length, they find that crack propagation may first happen in an unstable way, followed by a regime of stable growth.

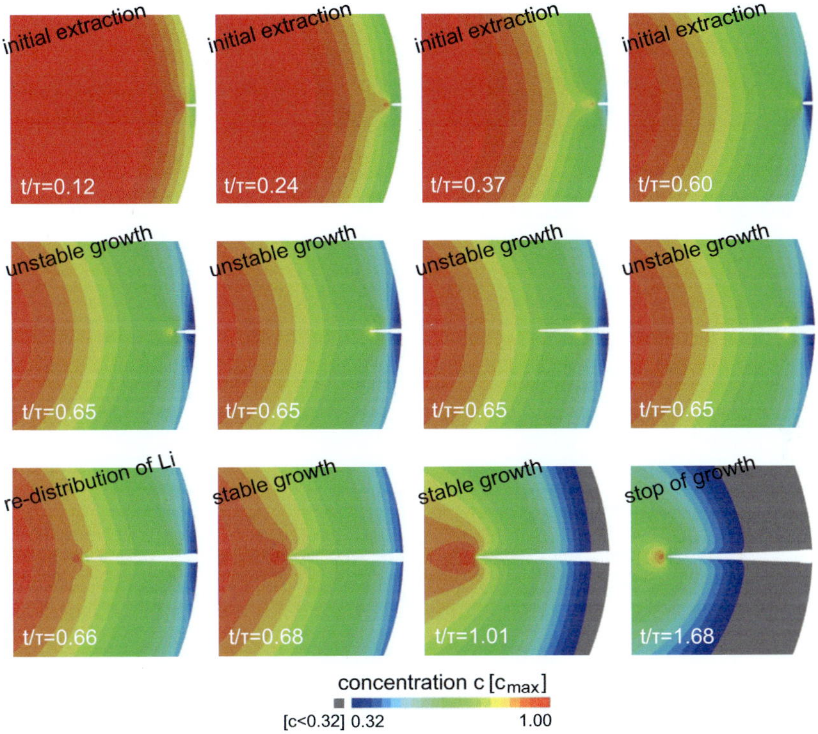

Figure 7.10.: Li concentration in units of maximum concentration $c_{\max}$ redistributing around the crack tip during crack growth in a storage particle with $r = 3\,\mu\text{m}$, $a = 0.1\,\mu\text{m}$ and $C = 10$. Regions in gray indicate values of the Li concentration $c < 0.32\,c_{\max}$. Times are given in units of $\tau = r_0^2/D_0$.

Effect of Crack Length

With respect to the considerations made at the beginning of this subsection, unstable crack growth should only occur when the initial crack length is small in comparison to the particle radius. This is reflected by the graph in FIG. 7.12, where we vary the initial crack length while keeping particle radius and C-rate constant at **r=5 µm** and **C=5**. We find that a critical initial crack length ($a_c \approx 0.06...0.07\,\mu\text{m}$) exists that must be exceeded in order to induce unstable crack growth for the chosen set of parameters. The extent of unstable crack growth then decreases with larger initial crack lengths a. For crack sizes larger than $a \gtrsim 0.5$, only stable crack growth takes place. As described in the explanation for FIG. 7.2, small extents of unstable crack growth are difficult to identify precisely in the simulations. In FIG. 7.12, the data point at $a = 0.4\,\mu$m therefore has to be considered as a rough estimate of the actual extent. Analysis of the time rate of change of the total crack surface area obtained in this case shows an estimated deviation of approximately $\pm 50\%$ compared to the plotted value of unstable growth extent. However, such a large error only applies to this data point.

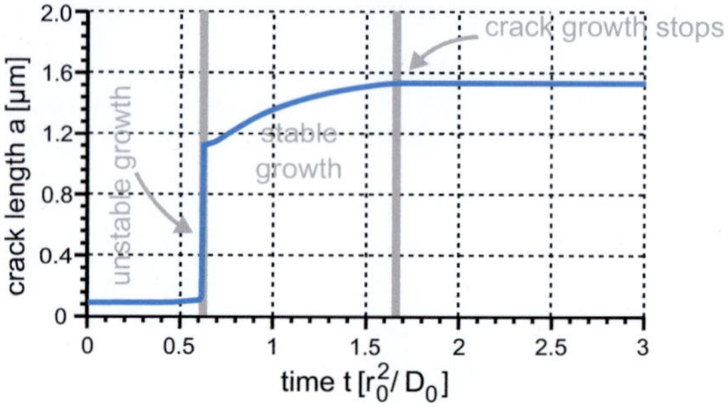

Figure 7.11.: Crack length during Li extraction in a storage particle with radius $r = 3\,\mu$m, initial crack length $a = 0.1\,\mu$m and $C = 10$. Unstable crack growth is followed by a regime of stable crack extension.

The dependence of the extent of unstable crack growth on the initial crack length, as illustrated in FIG. 7.12, is in agreement with the assessment made by means of FIG. 7.4 in section 7.1.1. For a given C-rate C and particle radius r, a certain initial crack length $\tilde{a}$ exists for which the stress at the crack tip is maximal. A smaller initial crack length $a < \tilde{a}$ results in less stress generated by the contracting material for radii $r' > r - a$. Larger values $a > \tilde{a}$ lead to a wider crack opening and less material within the radius $r' < r - a$, working against the contraction. Hence, the stress at the crack tip decreases. These two effects act counter to each other and only for a certain initial crack length $\tilde{a}$ is the tensile stress at the crack tip maximized with respect to a. For $a > \tilde{a}$, the stress decreases with growth of the crack and more Li needs to be extracted for further crack propagation to occur. Hence, crack growth is stable. If the initial crack length is too small, i.e. $a \leq a_c$, the stress at the crack tip is not high enough to induce crack growth. As in the case of Li insertion in chapter 6, the most pronounced unstable crack growth is therefore expected in the case where the crack length a is just slightly larger than the critical crack length a_c.

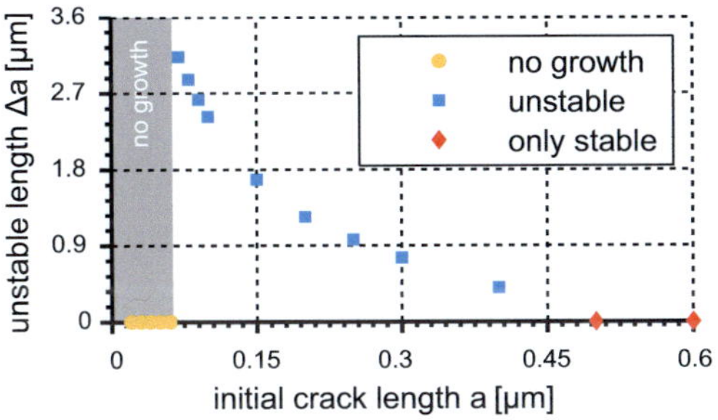

Figure 7.12.: Distance traveled by the crack tip during unstable crack growth for different values of the initial crack length a. The radius of the storage particle is $r = 5\,\mu$m and Li is extracted with $C = 5$.

Effect of C-rate

Following the findings in [182], the extent of unstable growth should not only depend on the initial crack length, but also on the flux boundary conditions. In FIG. 7.13, we therefore vary the C-rate, keeping the particle radius and initial crack length constant at **r=5 μm** and **a=0.1 μm**. In [182], the influence of an instantly applied surface temperature reduction on the behavior of crack growth was studied. In the case of Li concentration induced volume changes, this corresponds to an entirely potentiostatic Li extraction. Since the boundary condition applied in our simulation involves galvanostatic extraction followed by potentiostatic extraction, the results in FIG. 7.13 are not directly comparable to the ones in [182]. However, the fundamental trend is the same. First, a gradient in volume expansion above a threshold value is required to drive crack growth. In [182], this corresponds to a threshold in temperature difference between the surface and interior. Here, the gradient in volume expansion results from the inhomogeneous Li concentration and its magnitude is related to the C-rate. The minimum C-rate resulting in crack growth for the given particle radius r and initial crack length a is found to be $C_{\min} \approx 5$. When the gradient in volumetric expansion is increased by choice of smaller surface temperatures or higher C-rates, the extent of unstable crack growth decreases, as shown in FIG. 7.13. This relation is consistent with the mechanism described above on stress generation by shells of radius r', smaller and larger than $r - a$, and the influence of the crack opening. For large C-rates, the mechanical stress at the crack tip increases due to the stronger gradient in volumetric expansion. This shifts the critical crack length a_c to smaller values. For a C-rate of $C = 5$, the critical crack length is approximately equal to the crack length used in the simulation, $a = 0.1\,\mu\text{m}$. In this case, the C-rate is just sufficiently large to initiate crack growth. With further increase of the C-rate, the critical initial crack length a_c becomes smaller and the difference between the chosen initial crack length a and the critical one, a_c, increases. The distance traveled by the crack tip during unstable crack growth thus decreases.

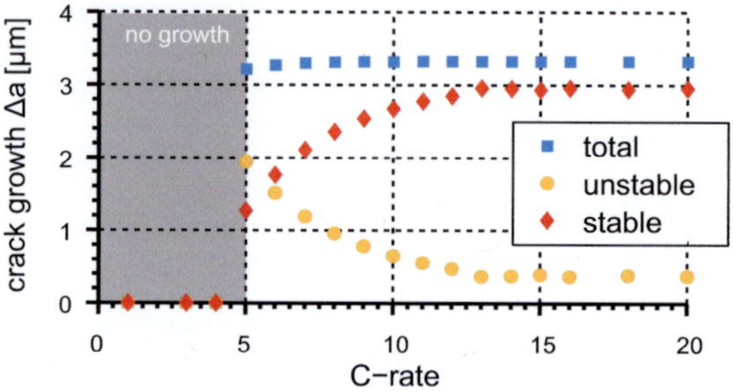

Figure 7.13.: Distance traveled by the crack tip during stable and unstable crack growth for different values of the C-rate C. The radius of the storage particle is $r = 5\,\mu\text{m}$ and the initial crack length is $a = 0.1\,\mu\text{m}$.

In [182], an upper critical temperature difference is determined above which only stable crack growth takes place. An analogous situation cannot be attained through variation of the C-rate for the given particle radius r and crack length a due to the following reason. With increasing C-rate, the extraction process resembles more closely an entirely potentiostatic Li extraction, as described in chapter 6. However, in contrast to the thermal model in [182], where the temperature difference between inner and outer part of the specimen can, in principle, take on arbitrarily high values, the maximum difference in Li concentration is bounded by the maximum Li concentration c_{max}. Consequently, as the C-rate increases, the extent of unstable crack growth converges to the value obtained for entirely potentiostatic Li extraction, with $c = 0$ at the surface.

Hence, for the particle size and initial crack length used, purely stable crack growth does not occur, independent of the selected C-rate.

Effect of Particle Radius

Besides initial crack length a and C-rate C, the particle radius r influences the occurrence of unstable crack growth as well. As illustrated in chapter 6, the particle radius has a strong impact on the minimum crack length and C-rate necessary to initiate crack growth. This can be explained by the characteristic time scale for Li to diffuse into the particle, which scales roughly with $\tau_r = r^2/D_0$. As the radius of the storage particle is increased, the time required for Li to equalize grows and larger differences of Li concentration build up. The resulting inhomogeneous volume expansion leads to higher stress in the storage particle. This relationship also applies to extraction and is reflected in TAB. 7.1, which shows that growth of initial cracks with specific lengths during Li extraction with given C-rates only occurs when the size of the storage particle is above a certain threshold. For example, under Li extraction with $C = 1$, initial cracks of length between 0.1 and 1.0 μm only grow when the storage particle radius r exceeds 5 μm. In addition the conditions under which cracks grow only in a stable way is affected by the storage particle radius r. Compare, for example, the simulations with $a = 0.1\,\mu$m and $C = 5$ in TAB. 7.1. For $r = 5\,\mu$m and $r = 10\,\mu$m initial crack extension is unstable, while for $r = 15\,\mu$m and $r = 20\,\mu$m the entire crack propagation is stable. This is consistent with the finding when the C-rate was varied that higher concentration gradients lead to more limited regions of unstable growth.

We find that unstable crack growth also occurs during extraction of Li. However, it is typically less pronounced than during insertion in the sense that crack tip velocities large enough to induce crack branching or particle breakage are not encountered during Li extraction. Also, while increasing C-rates and particle radii lead to faster unstable crack propagation during Li insertion, such conditions reduce the regime of unstable crack growth in the case of Li extraction.

7.1.4 Additional Cracks

In our simulations with the highest C-rates and largest particle radii, additional cracks are nucleated at locations, where there are no initial cracks. This effect is observed for $\mathbf{r \geq 15\,\mu m}$ and $\mathbf{C \geq 5}$, as shown in TAB. 7.1. An example of the resulting fracture pattern is given in FIG. 7.14. The additional cracks are a result of the smooth approximation of the discrete cracks inherent in the phase field method. The stress relaxed by growth of the two initial cracks in the horizontal symmetry axis, which are introduced through the history field, is not sufficient to significantly lower the stress level on the remote perimeter of the storage particle. The remaining high stress then leads to the creation of cracks at locations where the order parameter d is initially zero. In this situation, the length scale parameter l behaves to some extent as a material parameter, which defines an effective initial crack length. However, as was shown in chapter 5, it is doubtful whether the initial crack length for defects modeled in such a manner can be obtained from a specified value of l in a precise way. Further, with respect to the findings in chapter 4, predictions of the character of initial crack growth, i.e. stable or unstable, are likely to become less reliable for cracks modeled in this manner. Hence, the meaning of the results of the corresponding computations cannot be deduced in the same way as in simulations where the initial crack size is introduced in a well-controlled way. Therefore computations are terminated without evaluation of the results as soon as additional cracks, such as those depicted in FIG. 7.14, begin to grow.

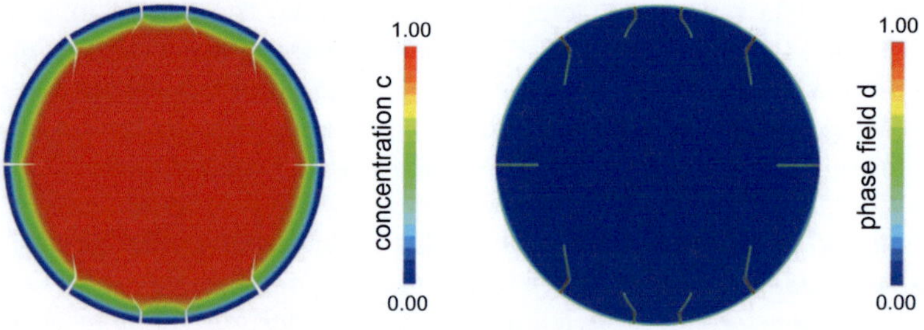

Figure 7.14.: Li extraction in a storage particle with $r = 10\,\mu\text{m}$, $a = 1\,\mu\text{m}$ and $C = 5$ after additional cracks have formed on its surface. (Left) Concentration c in units of maximum concentration c_{max}. Areas with $d > 0.95$ are removed to show the location of the crack. The opening of the crack is exaggerated by a linear scaling of the nodal displacements. (Right) Order parameter d at the same time step.

7.2 Spherical Particles in 3D

In order to study whether the results obtained in the 2D plane strain geometry are transferable to three-dimensional storage particles and to examine whether further phenomena related to the additional dimension exist, we run a representative number of three-dimensional simulations.

As in the 3D simulations of the study of crack growth during Li insertion in chapter 6, the length scale parameter is increased to $l = 0.25\,\mu\text{m}$, allowing a slightly coarser finite element mesh. This saves computation time but is still sufficiently small to resolve the initial cracks. The mesh adaptation scheme again generates a suitable mesh element size close to the crack faces, which is $h \leq (2/5)\,l$. Due to the stronger smoothing of the crack, the refined mesh stays below approximately 0.4 million in terms of number of elements or approximately 21 million in terms of degrees of freedom. However, simulations of crack growth during Li extraction typically take more time than those of Li insertion for the following reasons. First, since most of the crack growth during Li extraction is stable, the diffusion and fracture process take place on a similar time scale. In contrast to unstable crack growth, this means that during growth of the crack the concentration distribution changes by a significant amount. This results in a modification of the volume expansion due to the intercalated Li. For this reason, the difference between the displacement fields of two subsequent time steps is larger than during entirely unstable crack growth, where the concentration distribution remains approximately constant[1]. Hence, at each new time step, the displacement field is further away from the current solution. This slows down solving the non-linear equation describing conservation of linear momentum since more iterations of the Newton-Raphson method are required. Second, the large crack opening at the surface makes the stiffness matrix less well-conditioned due to the spectral decomposition of the strain tensor. For this reason, we increase the residual stiffness to $k = 5 \times 10^{-4}$ and focus on only one

[1] The Li concentration enters the equation for balance of linear momentum (2.7) through the constitutive equation for the stress (2.34) and the definition of the Li strain (2.10). In order to satisfy equation (2.7), the small strain tensor $\boldsymbol{\varepsilon}$, defined in equation (2.8), needs to compensate for a change in the Li strain. The displacement field thus experiences stronger changes when the concentration is altered by larger amounts. This is most pronounced close to the surface, where the integrated local volumetric expansion determines the total shrinkage of the particle. In the Li extraction study, this is also where the initial crack is located and where the equation for balance of linear momentum is strongly non-linear due to the decomposition of the strain tensor.

three-dimensional simulation with crack growth, as shown in the last row of TAB. 7.2. A second simulation, running with $k = 1 \times 10^{-5}$ and computing only an eighth of the sphere, showed qualitatively identical results and only small deviations in the values of computed parameters such as, for example, the hydrostatic stress.

Parameters			crack behavior	
r	a	C	growth	unstable
5	1	1	no	no
5	1	5	no	no
5	1	10	yes	yes

Table 7.2.: Input parameters, i.e. particle radius r, crack length a and C-rate C, as well as the type of crack growth, i.e. entirely stable or with an initial unstable phase, in the 3D extraction study.

7.2.1 Safe Conditions without Crack Growth

As in the two-dimensional study, we first consider a situation without crack growth. This is given, for example, by the parameter combination **r=5 μm, a=1 μm** and **C=5**. The corresponding phase field representation of the initial crack is shown on the right of FIG. 7.15.

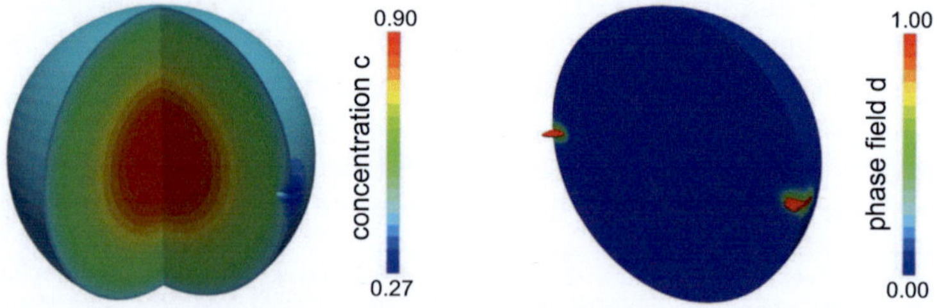

Figure 7.15.: (Left) Concentration c in units of maximum concentration $c_{\max}$ at an intermediate stage of Li extraction in a spherical storage particle with $r = 5\,\mu$m, $a = 1\,\mu$m and $C = 5$. (Right) Order parameter d at time $t = 0$. The quarter-disks protruding from the half sphere illustrate the volumes characterized by $d > 0.95$.

The graphs of results for the characteristic parameters plotted in FIG. 7.16 are similar to the ones obtained in section 7.1 for the cylindrical particle under plane strain. The state of charge first decreases linearly until the transition from galvanostatic to potentionstatic Li extraction is reached at $t \approx 3.9\,\tau$. As before, the difference in Li concentration is largest between the center of the particle and the material adjacent to the crack faces, where the lower hydrostatic stress repulses Li. This is illustrated in the left of FIG. 7.15. Due to the concentration dependent prefactor in the mechanically driven part of the flux, given in equation (2.20), the repulsive effect is stronger at earlier times when the Li concentration close to the surface is still above zero. However, high magnitudes of mechanical stress are not determined by small, localized volumes of material with low Li concentration, but instead by differences of the Li concentration between large volumes in the inner and outer part of the particle. Thus, the highest value of the maximum principal stress is found at the transition from galvanostatic to potentiostatic Li extraction, when the difference between the average Li concentration at the surface and the Li concentration in the center of the storage particle is largest.

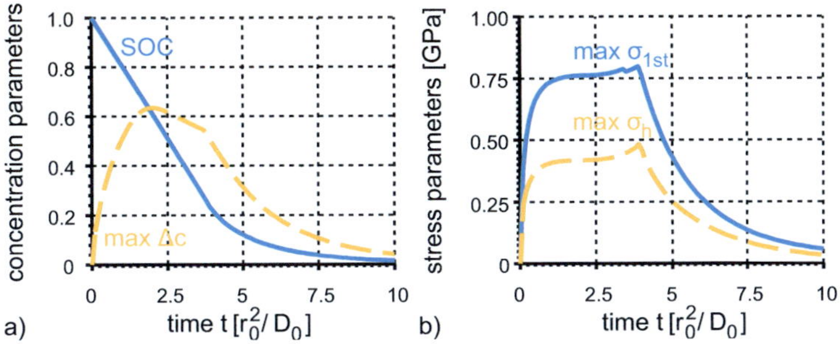

Figure 7.16.: Characteristic results during Li extraction in a 3D spherical particle with $r = 5\,\mu\text{m}$, $a = 1\,\mu\text{m}$ and $C = 5$. (a) State of charge (SOC) and maximum concentration difference $\max \Delta c$ in units of maximum concentration c_{max}. (b) The highest value of the maximum principal stress $\max \sigma_{1\text{st}}$ and maximum hydrostatic stress $\max \sigma_h$ during Li deintercalation.

We find that the highest value of the first principal stress ($\max \sigma_{1\text{st}} \approx 0.79\,\text{GPa}$) is only slightly larger than the highest value of the second principal stress ($\max \sigma_{2\text{nd}} \approx 0.72\,\text{GPa}$) at this point in time. This can also be observed by comparison of the first and second contour plot in

FIG. 7.17. The region around the crack, where the first principal stress acts perpendicular to the crack faces, is drawn in almost the same deep red color as the rest of the surface. At some distance away from the crack, the surface of the storage particles showing contour plots of the first and second principal stress, are colored in the same red tone. At these locations, the stress state is biaxial with a large tensile hoop stress and negligible ($\max \sigma_{1st} < 10\,\text{MPa}$) radial stress. In the center of the particle the stress state is triaxial and entirely compressive.

A small crack on the surface of the storage particle therefore influences the stress distribution within the spherical particle only by a small degree. This is similar to the case of Li insertion in chapter 6. However, stress directions during Li extraction are reversed with respect to Li insertion.

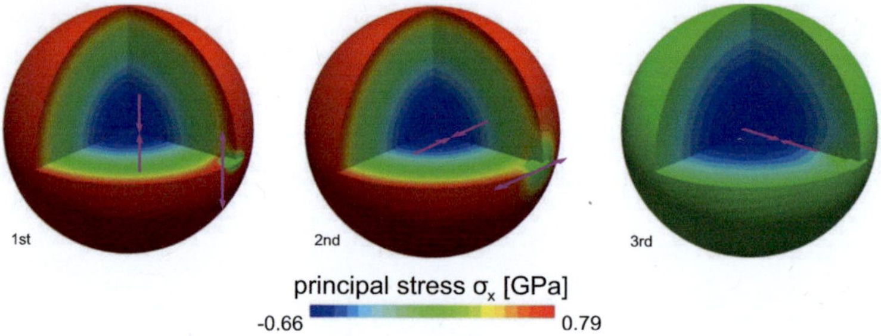

Figure 7.17.: Magnitude and orientation (pink arrows) of principal stresses at transition from galvanostatic to potentiostatic Li extraction in a storage particle with $r = 5\,\mu\text{m}$, $a = 1\,\mu\text{m}$ and $C = 5$.

7.2.2 Crack Growth

We increase the C-rate and examine the parameter combination $\mathbf{r{=}5\,\mu m}$, $\mathbf{a{=}1\,\mu m}$ and **C=10**. Due to the higher C-rate sufficient mechanical stress is generated to provoke growth of the initial crack. This can be observed by means of FIG. 7.18, which shows results for characteristic parameters involved in the Li extraction process.

The initiation of crack growth is marked by the first peak and subsequent drop of the highest value of the first principal stress $\max \sigma_{1st}$. The crack first extends along the surface of the particle in an unstable way, as shown in FIG. 7.19. During growth the crack front remains in the plane of the initial crack. The regime of unstable crack growth ends after the crack faces of the two initial cracks merge at around $t \approx 0.522\,\tau$ and a ring-like crack topology is created. This is illustrated in the second to last image in FIG. 7.19. Before and during unstable crack propagation, the largest tensile stress occurs at the segment of the crack front that is located at the particle surface. For this reason, the graphs in FIG. 7.18, showing the highest values of the maximum principal stress, found in the entire storage particle, at its surface and at the crack front, coincide with each other.

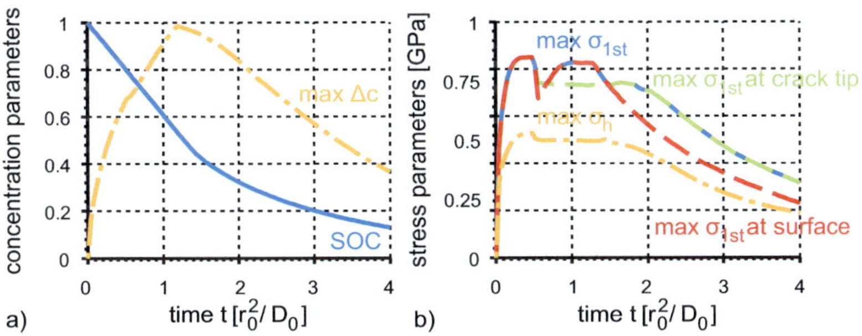

Figure 7.18.: Characteristic results during Li extraction in a 3D spherical particle with $r = 5\,\mu$m, $a = 1\,\mu$m and $C = 10$. (a) State of charge (SOC) and maximum concentration difference $\max \Delta c$ in units of maximum concentration $c_{\max}$. (b) The highest value of the hydrostatic stress $\max \sigma_h$ and the maximum principal stress $\max \sigma_{1\text{st}}$ in the entire particle and at surface and crack tip during Li deintercalation.

After completion of the formation of the ring crack, it extends further into the interior of the particle, now in a stable way. While the highest value of the maximum principal stress at the crack front remains constant, the stress at the surface rises again. The magnitude of the highest stress at the surface then surpasses that of the stress at the crack tip, which occurs at $t \approx 0.68\,\tau$. The difference in the evolution of the stress at the crack front and at the surface results from the shape and orientation of the crack. The increase of the crack opening relaxes only tensile stress in the direction perpendicular to the crack faces. Hence, the tensile hoop stress acting in the direction parallel to the ring crack is not lowered. The location of the highest value of the maximum principal stress therefore jumps from the crack front to the perimeter of the storage particle. The direction of the principal stress is also illustrated by the pink arrows on the right of FIG. 7.20.

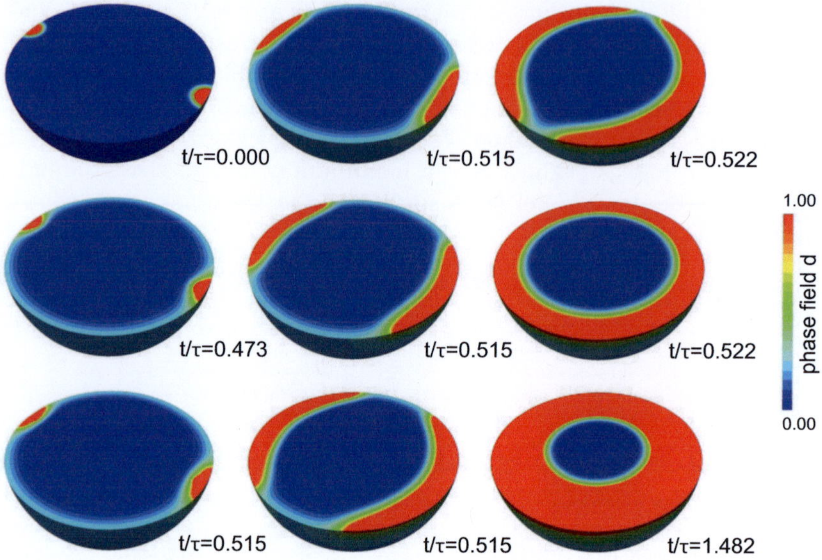

Figure 7.19.: Order parameter d during Li extraction and crack growth in a storage particle with $r = 5\,\mu\mathrm{m}$, $a = 1\,\mu\mathrm{m}$ and $C = 10$. Times are given in units of $\tau = r_0^2/D_0$. To improve the visibility of the progressing crack growth only the lower half of the storage particle is shown.

After galvanostatic Li extraction ends at $t \approx 1.28\,\tau$, the maximum difference in Li concentration in the storage particle decays. This results in a decrease of the highest value of the maximum principal stress. The reduction is fastest at the perimeter of the storage particle and at $t \approx 1.48\,\tau$, the location of the highest value of the maximum principal stress shifts back to the crack front. At this time, crack growth stops. However, as depicted by the green curve in FIG. 7.18, the stress at the crack tip is not relaxed as fast as at the perimeter of the storage particle. This can be explained by the dependence of the mechanical stress on the Li distribution within the storage particle. As described in section 7.1.1, the stress generated at the crack tip mainly depends on the overall contrast in volume change of the material located within radius $r' < r - a$ and $r' > r - a$, which results from the inhomogeneous Li concentration. Only to a lesser extent is it affected by the detailed distribution of Li within the various parts of the storage particle. This is different for the mechanical stress at the perimeter of the storage particle, which is strongly influenced by the gradient of Li concentration in the outermost shell of the storage particle. As the boundary condition shifts from galvanostatic to potentiostatic Li extraction, this gradient diminishes within a comparatively short time scale. In contrast, the impact of the transition to potentiostatic extraction on the evolution of the Li concentration in the interior of the storage particle requires some time to take effect. Hence, the average Li concentration in the volumes within radius $r' < r - a$ and $r' > r - a$ does not change as rapidly. As a consequence, the decrease of the stress at the crack tip is delayed with respect to that at the perimeter of the storage particle.

From these observations we conclude that the trajectory taken by the crack front during growth is highly dependent on the prevailing symmetry of the particle. Although the initial cracks disturb the rotational symmetry around the vertical center axes, the symmetry is restored by the initial pattern of crack growth. This first phase of crack growth takes place in an unstable way along the surface and is followed by stable extension of the crack towards the center of the storage particle. The surface crack does not fully relax the tensile stress on the surface of the storage particle. For this reason, additional initial surface cracks would be expected to grow, even though the first crack has significantly expanded in size.

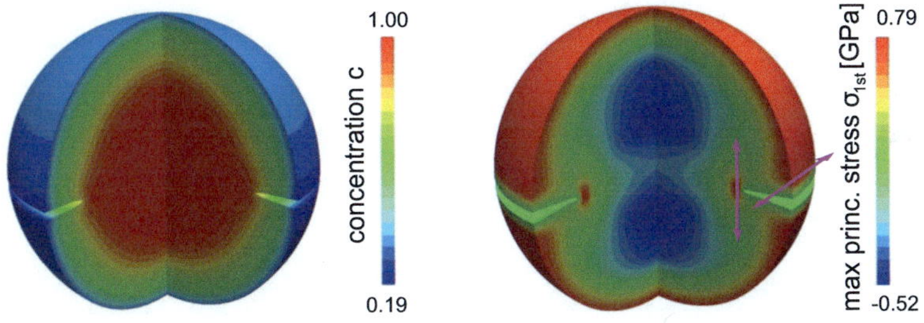

Figure 7.20.: (Left) Concentration c in units of maximum concentration c_{max} during Li extraction in a spherical storage particle with $r = 5\,\mu\text{m}$, $a = 1\,\mu\text{m}$ and $C = 10$. The time selected for the plots is $t \approx 0.8\,\tau$, with $\tau = r_0^2/D_0$. (Right) Highest value of the maximum principal stress $\sigma_{1\text{st}}$ at the same time. The pink arrows illustrate the local orientation of the maximum principal stress.

8 Crack Growth in the Second Half Cycle

In the study of crack growth during Li insertion in chapter 6, we saw that storage particles can break into two or several fragments when inertia forces, appearing during unstable crack growth, become strong enough to drive crack growth to the perimeter of the storage particle. This was not observed in the investigation on crack growth during Li extraction, described in chapter 7, where the regime of unstable crack growth is typically less pronounced. However, by comparison of TAB. 6.2 and 7.1 we observe that the critical conditions required for crack growth, i.e. the minimum crack size, particle radius and C-rate, are generally lower for a surface crack during Li extraction. This can be understood from the following simple argument. In the case of Li extraction in a storage particle with an initial surface crack, the crack opens to one side. Hence, the mechanical force, tearing the crack faces apart, only acts on one crack tip. In the case of Li insertion with an initial through crack, the critical stress required to induce crack growth has to be reached at two crack tips. Consequently, a higher total force perpendicular to the crack faces is necessary. This poses the interesting question of whether cracks, that have grown during Li extraction, are prone to propagate further during the next half cycle of Li insertion and whether particle breakage may also occur in a multi-step process. In this case, the comparatively wide range of parameter combinations still avoiding breakage of storage particles, that was found in the study on crack growth during Li insertion, might be narrowed since the first step of crack growth may take place during Li extraction, where the limits to prevent crack growth are lower. At the same time, crack growth during Li insertion commencing from a large initial crack may not lead to breakage of the storage particle due to the less pronounced unstable crack growth in this scenario. However, if the crack tip has propagated into regions that experience tensile stress during Li extraction, the storage particle may nevertheless break in a subsequent half cycle. In this chapter, we thus investigate crack growth and Li diffusion in the second half cycle.

To examine the fracture behavior in the second half cycle, we take as initial conditions the final state of some of the simulations presented in chapter 6 and 7. The material and numerical parameters are adopted from the respective computations without changes. Further, the absolute value of the applied flux is the same as in the corresponding initial half cycle and only its direction is reversed. We do not attempt to investigate all parameter combinations, presented in TAB. 6.2 and 7.1, but instead select a subset that covers the two central scenarios, with and without crack growth in the second half cycle. Storage particles that break into parts during the first half cycle are not further examined.

8.1 Lithium Extraction with Initial Through Crack

8.1.1 Cylindrical Particles in 2D under Plane Strain

In the study on crack growth during Li extraction in chapter 7, it is found that the critical stress required to induce crack growth is more easily generated in storage particles with large radius and during Li extraction with a high C-rate. Further, during Li extraction, the strongest tensile stress occurs at the perimeter of the storage particle. Hence, cracks that end close to the surface of the storage particle are more prone to crack growth. Among the combinations of parameters, that do not lead to crack growth during Li insertion (see TAB. 6.2), the one with storage particle radius $\mathbf{r=5\,\mu m}$, initial crack length $\mathbf{a=2\,\mu m}$ and C-rate $\mathbf{C=1}$ is therefore the one with the highest probability to show crack growth during Li extraction in the second half cycle. In order to evaluate whether crack growth in the second half cycle takes place even if the storage particle resists crack growth during the first half cycle, we select this combination of parameters and simulate the process of Li extraction, taking as initial conditions the final state of the first half cycle. The corresponding characteristic parameters are plotted in FIG. 8.1.

The graphs showing the characteristic concentration and stress parameters are identical to those obtained in a simulation without an initial crack in the center of the storage particle. Hence, both the highest value of the maximum principal stress and the highest value of the hydrostatic stress is not located at the crack tip, but instead is found at the surface of the storage particle. This can be understood by the following argument. The hoop stress in a storage particle without crack is given by a thermal analogy in [171] as

$$\sigma_\theta(r') = \frac{\Omega E}{6(1-\nu)} \left(c_{\mathrm{av}}(r) + c_{\mathrm{av}}(r') - 2c(r')\right). \tag{8.1}$$

The average concentration within the radius $r' \in [0, r]$ of a storage particle with outer radius r is defined through $c_{\mathrm{av}}(r') = \frac{2}{r'} \int_0^{r'} c(\tilde{r}')\tilde{r}'\mathrm{d}\tilde{r}'$. Similar to the considerations in section 6.1.2, we assume a Li concentration distribution along the radius that is approximated by a power law $c(r') = (c_o - c_i)(r'/r)^\gamma + c_i$ with the concentrations c_i and c_o at the center and perimeter of the storage particle respectively. The Li content along the radius is typically concave during extraction, which requires an exponent of $\gamma > 1$ due to $c_0 < c_i$. The resulting hoop stress then reads

$$\sigma_\theta(r') = \frac{\Omega E}{3(1-\nu)} \frac{c_o - c_i}{\gamma + 2} (1 - (2(\gamma + 2) - 1)(r'/r)^\gamma). \tag{8.2}$$

For $r'/r \leq 1/5$, i.e. from the crack tip to the center of the storage particle, and $\gamma > 1$, this expression yields negative values of σ_θ. The same argument holds for the radial stress. Hence, both hoop and radial stress are compressive at the location of the crack. Consequently, the crack faces do not open and the material behaves as if there was no crack.

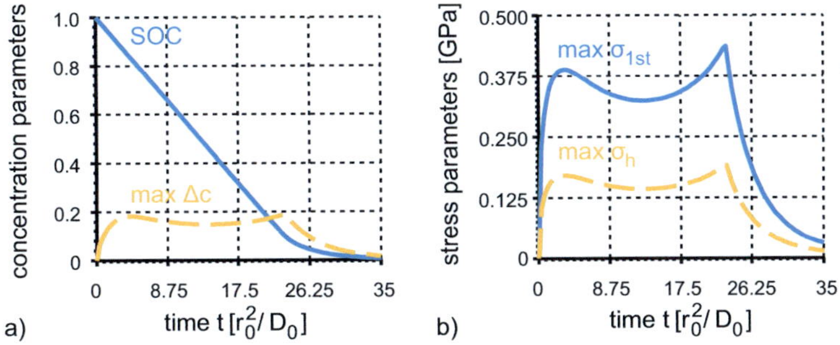

Figure 8.1.: Characteristic results for Li extraction in second half cycle with $r = 5\,\mu$m, $a = 2\,\mu$m and $C = 1$. (a) State of charge (SOC) and maximum concentration difference $\max \Delta c$ in units of maximum concentration $c_{\max}$. (b) The highest value of the maximum principal stress $\max \sigma_{1\mathrm{st}}$ and maximum hydrostatic stress $\max \sigma_h$ during Li extraction.

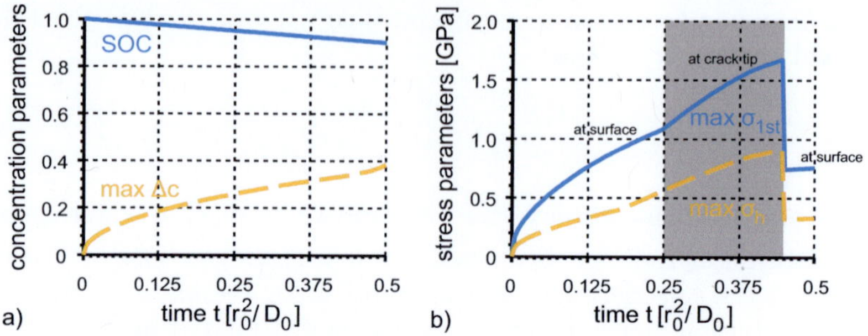

Figure 8.2.: Characteristic results for Li extraction in second half cycle with $r = 5\,\mu\text{m}$, $a = 2\,\mu\text{m}$ and $C = 5$. (a) State of charge (SOC) and maximum concentration difference $\max \Delta c$ in units of maximum concentration c_{max}. (b) The highest value of the maximum principal stress $\max \sigma_{1\text{st}}$ and maximum hydrostatic stress $\max \sigma_h$ during Li extraction.

In order to examine whether the situation changes when we consider Li extraction during the second half cycle in storage particles that already experience crack growth in the first half cycle, we repeat the procedure for a storage particle with the same radius and initial crack length, but with increased C-rate. The corresponding parameter combination is thus given by $\mathbf{r{=}5\,\mu m}$, $\mathbf{a{=}2\,\mu m}$ and $\mathbf{C{=}5}$. In this case, the crack growth during Li insertion in the first half cycle stops close to the perimeter of the storage particle, with an extension of $\Delta a \approx 3.86\,\mu\text{m}$ in one direction. The final diameter is $9.72\,\mu\text{m}$, as can be seen in FIG. 6.6 and in the left contour plot of FIG. 8.3. Due to the close proximity of crack tip and surface of the storage particle, equation (8.2) now predicts a tensile hoop stress at the crack tip during Li extraction. This is confirmed from analysis of the stress field, obtained in the simulation, and can also be observed indirectly by the attraction of Li on the left hand side of FIG. 8.2. The mechanical stress at the crack tip not only becomes tensile, but also larger than on the particle surface. This is marked by the gray region in FIG. 8.2. Further, the magnitude of tensile stress acquired at the crack tip becomes sufficiently large to initiate further crack growth. Since the hoop stress does not decrease with respect to the radial distance to the center of the storage particle and because less material has to keep the two halves of the particle together when the crack extends, the resulting crack growth takes place in an unstable way. This is indicated by the sharp drop of the stress parameters in FIG. 8.2 b). After the particle has broken into two, the largest tensile stress is again located at the perimeter of the storage particle. Due to the separation into two fragments, as depicted on the right of FIG. 8.2, the highest value of the maximum principal stress is now lower than before breakage.

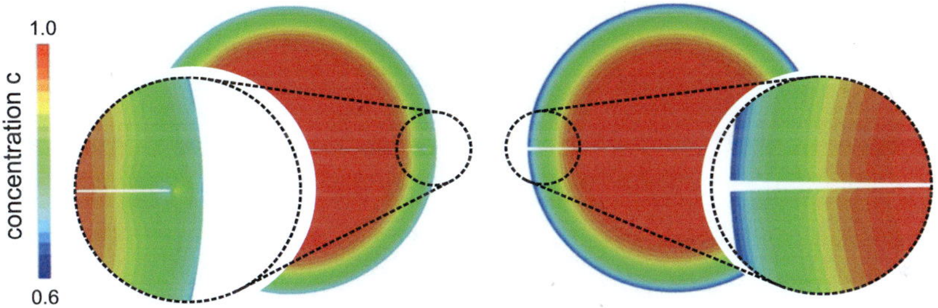

Figure 8.3.: Concentration c in units of maximum concentration $c_{\max}$ in a storage particle with $r = 5\,\mu$m, $a = 2\,\mu$m and $C = 5$, shown before ($t \approx 0.37\,\tau$) and after ($t \approx 0.49\,\tau$) breakage of the storage particle. Areas with $d > 0.95$ are removed to show the location of the crack. The opening of the crack is exaggerated by a linear scaling of the nodal displacements.

8.1.2 Spherical Particles in 3D

Due to the computational cost of the three-dimensional simulations, the choice of simulated Li insertion processes available as starting point for computation of the second half cycle is limited. However, the central cases, i.e. with and without crack growth during the second half cycle, are covered by this selection. Again, we first examine a combination of parameters that does not lead to crack growth during Li insertion in the first half cycle. This is given by **r=5 μm**, **a=1 μm** and **C=5**.

As in the two-dimensional case, the graphs of the characteristic concentration and stress parameters, shown in FIG. 8.4, during Li extraction in the second half cycle in a storage particle, that does not experience crack growth in the previous half cycle, cannot be distinguished from the case of Li extraction in a storage particle without cracks. As before, this can be explained by means of analytic relations of the type given in equation (8.1) and (8.2). The different geometries of the two- and three-dimensional cases only affect the corresponding relations by prefactors.

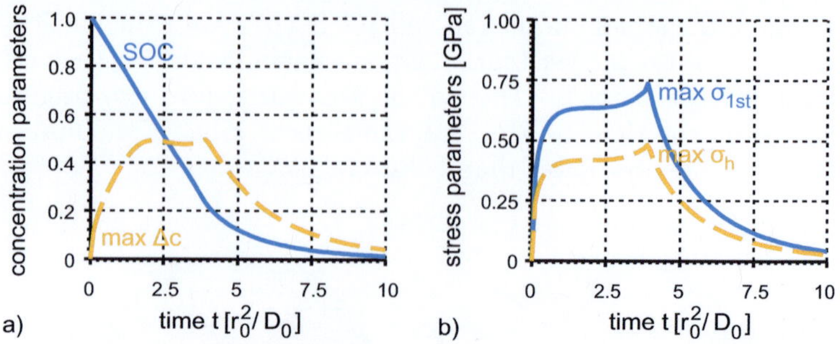

Figure 8.4.: Characteristic results for Li extraction in second half cycle with $r = 5\,\mu\mathrm{m}$, $a = 1\,\mu\mathrm{m}$ and $C = 5$. (a) State of charge (SOC) and maximum concentration difference $\max \Delta c$ in units of maximum concentration $c_{\max}$. (b) The highest value of the maximum principal stress $\max \sigma_{1\mathrm{st}}$ and maximum hydrostatic stress $\max \sigma_h$ during Li extraction.

Hence, assuming the same concentration distribution along the radial coordinate, we obtain the hoop stress at distance r' from the center by

$$\sigma_\theta(r') = \frac{\Omega E}{3(1-\nu)} \frac{c_o - c_i}{\gamma + 3} (2 - (\gamma + 2)(r'/r)^\gamma). \tag{8.3}$$

For $\gamma > 1$ this predicts compressive stress within a radius of $r'/r \leq 2/3$, which is well above the crack extension of $a/r = 1/10$. Hence, the entire crack is again under compressive stress and the mechanical behavior of the material is not affected by the presence of the crack. Comparing the radial distance at which equations (8.2) and (8.3) yield zero hoop stress for the same value of γ, we find that the region of compressive stress in the three-dimensional spherical particle is significantly larger than in the two-dimensional cylindrical particle under plane strain. For $\gamma = 1$, the radial coordinate of zero hoop stress is $r'/r = 2/3$ and $r'/r = 1/5$ respectively. The difference results from the different volumes V_{shell}, occupied by shells with the same thickness and distance to the center of the storage particle, in a sphere and in a cylinder[1]. In the case of the three-dimensional sphere, the volume of the shell scales as $V_{\mathrm{shell}} \propto r'^2$. For the two-dimensional cylinder, it goes with $V_{\mathrm{shell}} \propto r'$. During Li

[1] For the illustration of such a shell, see the region in orange color, shown in FIG. 7.4, with omission of the crack.

extraction, the outer shell contracts, resulting in a compressive force on the inner part of the storage particle. The stress in the storage particle is determined by a balance of force between this outer shell and the inner core. Since the volume of the outer shell in a spherical storage particle grows faster with respect to the radial distance to the center than in the cylindrical particle, the radius at which hoop stress becomes tensile is shifted closer to the surface. Hence, in order to experience tensile stress during Li extraction, the crack tip of a through crack needs to be located closer to the perimeter of the storage particle.

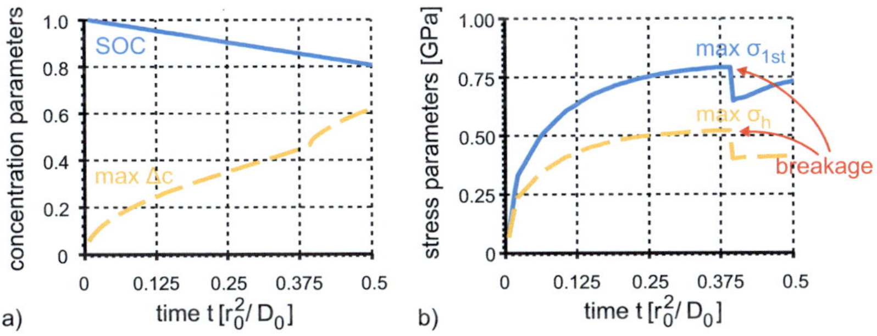

Figure 8.5.: Characteristic results for Li extraction in second half cycle with $r = 5\,\mu\text{m}$, $a = 2\,\mu\text{m}$ and $C = 10$. (a) State of charge (SOC) and maximum concentration difference $\max \Delta c$ in units of maximum concentration c_{max}. (b) The highest value of the maximum principal stress $\max \sigma_{1\text{st}}$ and maximum hydrostatic stress $\max \sigma_h$ during Li extraction.

The three-dimensional simulation of crack growth during Li insertion showing crack growth without breakage of the storage particle, is given by the parameter combination **r=5 μm, a=2 μm** and **C=10**. The final diameter of the crack after the first half cycle is approximately $8.9\,\mu\text{m}$, which corresponds to a radial extension of the crack of $\Delta a \approx 3.45\,\mu\text{m}$. The material behavior during the second half cycle of Li extraction is similar to the case with crack growth in the two-dimensional geometry. In the initial phase of Li extraction, highest tensile stress occurs at the surface of the storage particle. The location of highest tensile stress then switches to the crack tip, where it becomes large enough to initiate crack growth. The transition of the location of the highest value of the maximum principal stress does not affect the graphs of the stress parameters, plotted in FIG. 8.5, as noticeably as in the two-dimensional setting, shown in FIG. 8.2. However, the transition is confirmed by analysis of the three-dimensional stress distribution in contour plots during post-processing.

As soon as the critical stress for crack growth is reached, the storage particle breaks into two almost instantly. This is marked by the sharp drop in the stress parameters at $t \approx 0.39\,\tau$. After breakage of the storage particle, the highest tensile stress is again located at the particle surface. Its magnitude is now slightly lower than before breakage of the storage particle. The reduction of the tensile stress is a consequence of the deformation of the storage particle due to the tensile stress at its surface and the free motion of the crack faces after breakage. In order to illustrate this effect more clearly, the deformation of the storage particle, displayed on the left of FIG. 8.6, is scaled by a factor of 10.

Both the two- and three-dimensional simulations, as well as analytic relations, show that crack growth during Li extraction starting from an initial through crack in the center of the storage particle only occurs if the crack tip is located close to the perimeter of the particle. This is typically the case if the crack has already grown in the previous half cycle. If crack growth occurs in the second half cycle, it is unstable and leads to breakage of the storage particle. This also confirms the hypothesis on page 134 of section 6.2.4 that small ligaments at the surface, holding a particle together, cannot withstand the high stress in the following half cycle.

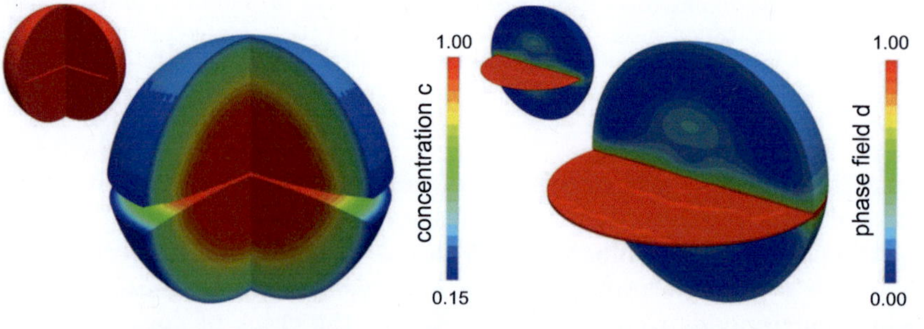

Figure 8.6.: (Left) Concentration c in units of maximum concentration $c_{\max}$ after the storage particle has broken in two. Areas with $d > 0.95$ are removed to show the location of the crack. The opening of the crack is exaggerated by a linear scaling of the nodal displacements. (Right) Order parameter d at the same time step. The protruding volume marks the region with $d \geq 0.95$. The inlets show the concentration c and order parameter d at the beginning of the second half cycle.

8.2 Lithium Insertion with Initial Surface Crack

8.2.1 Cylindrical Particles in 2D under Plane Strain

With respect to the finding from the studies on crack growth during Li insertion and Li extraction, presented in chapter 6 and 7 respectively, it is known that crack growth during Li insertion typically requires more severe conditions than during Li extraction, i.e. a larger storage particle radius and crack size and a higher C-rate. Since crack growth during Li extraction in the second half cycle is not observed if the crack has not undergone growth in the first half cycle of Li insertion, it is therefore also not expected in the corresponding scenario with swapped order of Li extraction and insertion. For this reason, we choose a parameter combination that already shows crack growth during the first half cycle. The selected combination of parameters is given by **r=5 μm**, **a=0.1 μm** and **C=5**. The individual extension of the cracks obtained in the respective first half cycle is $\Delta a \approx 3.3\,\mu$m.

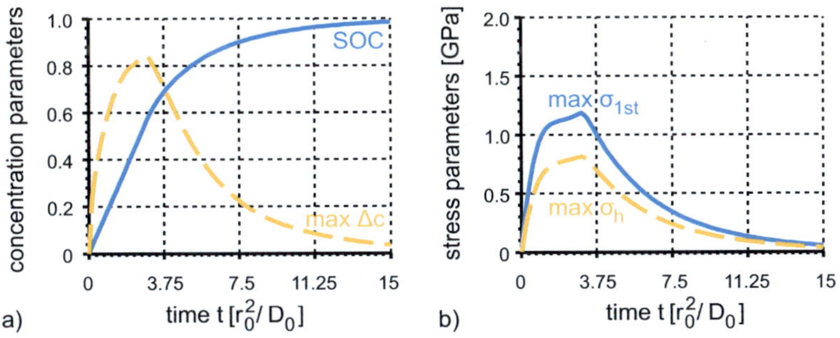

Figure 8.7.: Characteristic results for Li insertion in second half cycle with $r = 5\,\mu$m, $a = 0.1\,\mu$m and $C = 5$. (a) State of charge (SOC) and maximum concentration difference $\max \Delta c$ in units of maximum concentration c_{max}. (b) The highest value of the maximum principal stress $\max \sigma_{\text{1st}}$ and maximum hydrostatic stress $\max \sigma_h$ during Li intercalation.

From the graphs of the characteristic concentration and stress parameters, plotted in FIG. 8.7, we can deduce that further crack growth does not occur under these circumstances. The highest value of the maximum principal stress $\max \sigma_{\text{1st}}$ grows monotonically until the transition to potentiostatic insertion is reached at $t \approx 3\,\tau$. At this point, also the maximum concentration difference $\max \Delta c$ is largest. A drop in the characteristic stress parameters indicating crack growth is not observable. This is confirmed by analysis of the final state of the phase field, which is identical to the initial one.

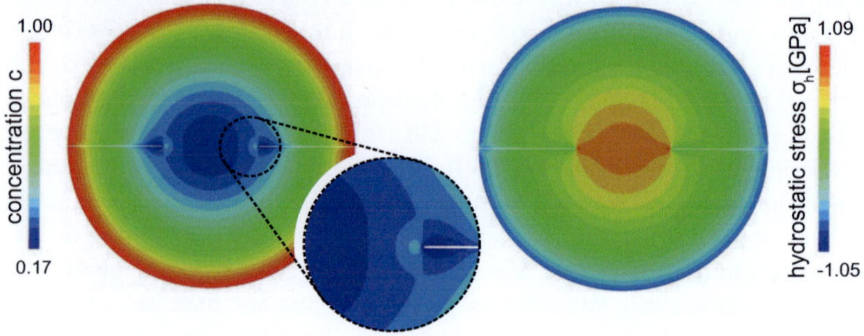

Figure 8.8.: (Left) Concentration c in units of maximum concentration c_{max} in a storage particle with $r = 5\,\mu\text{m}$, $a = 0.1\,\mu\text{m}$ during Li insertion in the second half cycle with $C = 5$, taken at $t = 2.98\,\tau$. (Right) The hydrostatic stress σ_h at the same time step. Areas with $d > 0.95$ are removed to show the location of the crack. The opening of the crack is exaggerated by a linear scaling of the nodal displacements.

Although the crack does not grow in the second half cycle, its presence influences the concentration and stress evolution in a noticeable way. This is in contrast to the investigation on Li extraction in the second half cycle, presented in section 8.1.1, where the stress and Li concentration distribution in the storage particle with initial through crack is identical to that without any crack. The effect of the crack on the stress and Li concentration distribution is illustrated in FIG. 8.8. We observe that the highest tensile stress occurs at the crack tip. This tensile stress attracts Li directly at the crack tip. However, in the material adjacent to the crack faces close to the crack tip, Li is repulsed due to the comparatively low hydrostatic stress. Hence, the maximum concentration difference close to the transition from galvanostatic to potentiostatic Li insertion at $t = \tau_{\text{galvano}}$ takes place between the perimeter of the storage particle and the material at the crack faces close to the crack tip. The interplay of Li diffusion and mechanical stress leads to a maximum concentration difference $\max \Delta c$ that seems to converge to a value below one before potentiostatic insertion is reached, as shown in FIG. 8.9. The decreasing slope of $\max \Delta c$ is reflected by a slower rise of $\max \sigma_{1\text{st}}$.

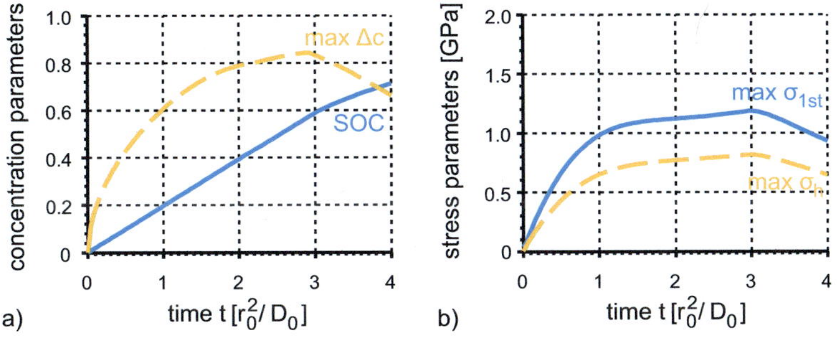

Figure 8.9.: Magnified view on graphs shown in FIG. 8.7. (a) State of charge (SOC) and maximum concentration difference max Δc in units of maximum concentration $c_{\max}$. (b) The highest value of the maximum principal stress max $\sigma_{1\text{st}}$ and maximum hydrostatic stress max σ_h during Li intercalation.

Although the critical stress required to initiate crack growth is not acquired, the highest value of the maximum principal stress of max $\sigma_{1\text{st}} \approx 1.2$ GPa, reached at $t = \tau_{\text{galvano}}$, is in the range of the typical magnitude observed during crack growth of max $\sigma_{1\text{st}} \approx 1.68$ GPa, as, for example, shown in FIG. 8.2. Hence, we expect that a slight increase of the radius of the storage particle or the C-rate is sufficient to generate enough stress to induce further crack growth. We therefore investigate the same combination of parameters, but with a higher C-rate, i.e. **r=5 μm, a=0.1 μm** and **C=10**.

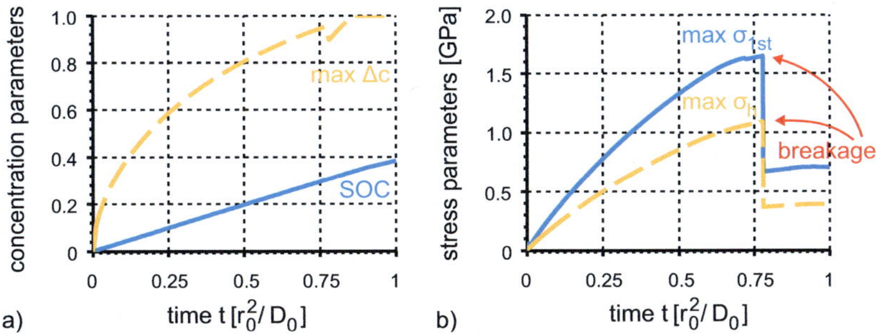

Figure 8.10.: Characteristic results for Li insertion in second half cycle with $r = 5\,\mu$m, $a = 0.1\,\mu$m and $C = 10$. (a) State of charge (SOC) and maximum concentration difference max Δc in units of maximum concentration $c_{\max}$. (b) The highest value of the maximum principal stress max $\sigma_{1\text{st}}$ and maximum hydrostatic stress max σ_h during Li intercalation.

With respect to the graphs of the characteristic concentration and stress parameters plotted in FIG. 8.10, we find that crack growth is initiated and that it leads to complete breakage of the storage particle. The incident of particle breakage can be observed by the sharp drop in the graphs of the stress parameters at $t \approx 0.78\,\tau$. As in the study on crack growth during Li extraction in the second half cycle, the sharp decrease of the stress parameters indicates that crack growth takes place in an unstable way. This results from the increase of tensile stress with smaller distance to the center of the storage particle and the decrease of material holding the upper and lower half of the storage particle together. An indication for the higher stress in the center is given by the larger crack opening displacement at $r' = 0$ in comparison to that at $r' = r$, as depicted in FIG. 8.11. Thus, as soon as crack growth is initiated in the second half cycle, the particle breaks into parts.

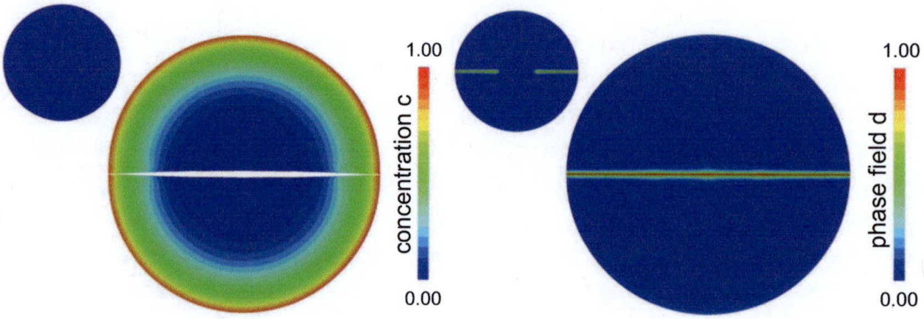

Figure 8.11.: (Left) Concentration c in units of maximum concentration c_{max} in a storage particle with $r = 5\,\mu\text{m}$, $a = 0.1\,\mu\text{m}$ during Li insertion in the second half cycle with $C = 10$, shown after breakage of the storage particle. Areas with $d > 0.95$ are removed to show the location of the crack. The opening of the crack is exaggerated by a linear scaling of the nodal displacements. (Right) The order parameter d at the same time step.

8.2.2 Spherical Particles in 3D

In order to examine possible differences between the fracture behavior of cylindrical particles under plane strain and spherical particles during Li insertion in the second half cycle, we perform two three-dimensional simulations. The respective starting points for the simulations are final states of computations presented in section 7.2. As before, we commence with a parameter combination that does not lead to crack growth in the first half cycle. This is given by **r=5 μm**, **a=1 μm** and **C=5**. The characteristic parameters during Li insertion are plotted in FIG. 8.12.

The graphs describing the time history of the characteristic Li concentration and stress parameters are identical to those obtained in an intact storage particle without initial crack. This is most clearly observable from the congruent graphs of the highest value of the maximum principal and hydrostatic stress. The stress state in the center of the storage particle, where tensile stress is largest during Li insertion, is not influenced by the cracks at the surface. Hence, even though the rotational symmetry around the vertical axis of the storage particle is broken due to the presence of the cracks, the stress state at the center is triaxial with equal magnitude in all directions. For this reason, the highest value of the maximum principal and hydrostatic stress coincide with each other. The vanishing influence of the cracks on the stress distribution results from the purely compressive stress state in those parts of the storage particle, where the cracks are located. Since the cracks do not extend far enough into to particle to experience tensile stress, the cracks do not open and the material behaves as if it was intact. This can also be explained by analytic relations of the type presented in equation (8.3).

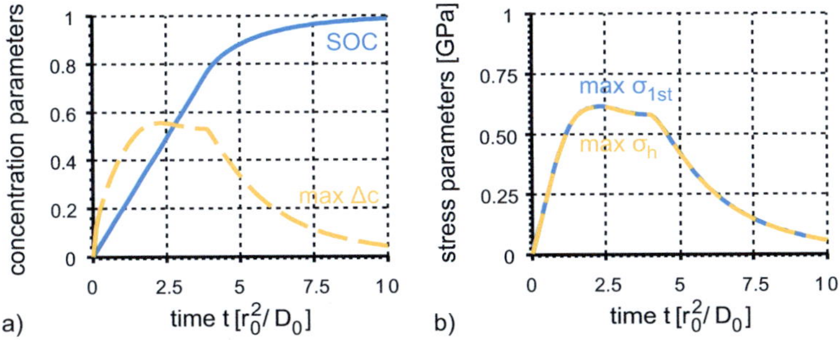

Figure 8.12.: Characteristic results for Li insertion in second half cycle with $r = 5\,\mu$m, $a = 1\,\mu$m and $C = 5$. (a) State of charge (SOC) and maximum concentration difference max Δc in units of maximum concentration $c_{\max}$. (b) The highest value of the maximum principal stress max $\sigma_{1\text{st}}$ and maximum hydrostatic stress max σ_h during Li intercalation.

Next, we investigate the parameter combination **r=5 μm**, **a=1 μm** and **C=10**. The corresponding simulation of the first half cycle is presented in section 7.2.2 and shows crack growth with an extension of $\Delta a \approx 1.6\,\mu$m of the crack front towards the center. The characteristic parameters observed during Li insertion in the second half cycle are shown in FIG. 8.13.

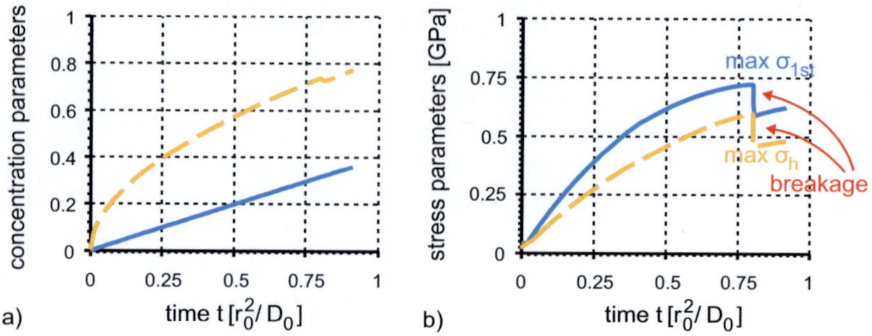

Figure 8.13.: Characteristic parameters during Li insertion in second half cycle with $r = 5\,\mu$m, $a = 1\,\mu$m and $C = 10$. (a) SOC and maximum concentration difference $\max \Delta c$ in units of maximum concentration $c_{\max}$ versus time in units of $\tau = r_0^2/D_0$. (b) Maximum first principal stress $\max \sigma_{1\text{st}}$ and maximum hydrostatic stress $\max \sigma_h$ during Li intercalation.

We observe that at around $t \approx 0.84\,\tau$, the highest value of the maximum principal stress drops from around $\max \sigma_{1\text{st}} = 7.26\,$GPa to $5.97\,$GPa, indicating that crack growth takes place in the second half cycle for the selected set of parameters. The crack growth is entirely unstable and leads to breakage of the particle into two halves, as illustrated in FIG. 8.14. In the moment of breakage of the storage particle, the location of the highest maximum principal stress jumps from the ring-like crack front to two symmetrically equivalent points on the vertical symmetry axis of the storage particle. The distance of these points to the center of the storage particle is approximately $z \approx 2.55\,\mu$m. This leads to a slight increase of the order parameter around the vertical symmetry axis close to the center of the storage particle, as illustrated by the image on the right of FIG. 8.14.

In comparison to the case with $C = 5$, the crack growth in the second half cycle results from two reasons. First, a higher C-rate leads to larger stress in the particle. And second, the crack growth during the first half cycle has promoted the crack front further into the particle. As was shown by the considerations on equation (8.3), the stress at the crack tip is highly dependent on its location in the particle. We therefore attribute the crack growth during Li insertion in the second half cycle mainly to the extension of the crack in the first half cycle.

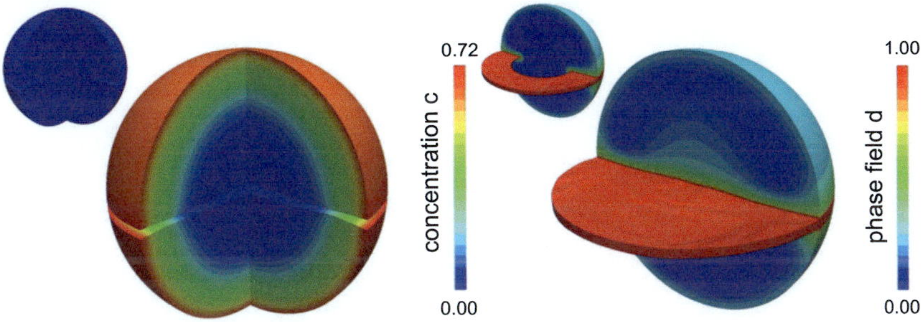

Figure 8.14.: (Left) Concentration c in units of maximum concentration c_{max} in a storage particle with $r = 5\,\mu\mathrm{m}$, $a = 1\,\mu\mathrm{m}$ during Li insertion in the second half cycle with $C = 10$, shown after breakage of the storage particle. Areas with $d > 0.95$ are removed to show the location of the crack. The opening of the crack is exaggerated by a linear scaling of the nodal displacements. (Right) The order parameter d at the same time step. The protruding volume marks the region with $d \geq 0.95$. The inlets show concentration and order parameter at the beginning of the second half cycle.

In contrast to Li insertion in the first half cycle, where crack growth is generally expected to result in breakage of the storage particle, either in the first half cycle or during subsequent extraction in a second half cycle (cf. section 8.1), extension of a surface crack during extraction is not a sufficient indicator for subsequent particle breakage. Although the critical conditions for growth of a surface crack are less severe, the extension of the crack is typically smaller than that of a through crack during Li insertion. Hence, the crack tip may come to rest in a region that does not experience sufficiently high stress to promote further crack growth in the following half cycle. However, if crack growth is initiated, it takes place in an unstable way and leads to fragmentation of the particle.

9 Lithium Diffusion at Crack

As described in the formulation of the model in chapter 2, a direct influence of the crack on the diffusion of Li within the storage particle is neglected in this work. The presence of the crack therefore influences the Li concentration distribution only through the modification of the hydrostatic stress. In the study on crack growth during Li insertion of chapter 6, we argue that this assumption is reasonable when the crack is closed, so that electrolyte cannot flow into it, and when either due to reasons of symmetry or because of the comparatively fast crack propagation, Li is not driven to cross the crack faces. In contrast, when the crack is located at the surface of the storage particle and opens to the surrounding, the crack might fill with electrolyte. This is the case in the investigation on crack growth during Li extraction, presented in chapter 7. Under these circumstances, Li could flow into or out of the crack through transport in the electrolyte and may intercalate or deintercalate at the crack faces. Further, in more generic geometries of storage particles than those examined in chapter 6 and 7, the argument of symmetry does not hold since the crack does not lie in a symmetry plane. In this scenario, the diffusion of Li across the crack could be blocked or at least hindered.

A physically more accurate description of the Li flux at the crack faces and the crack tip is beyond the scope of this work. Also, validation of more elaborated assumptions is difficult if not impossible at the current state since corresponding experimental observations are missing. However, the phase field approach for fracture provides a large flexibility[1]

[1] In addition, the complexity to implement modifications to the balance of Li, resulting from the diffusion at the crack, is facilitated by the staggered solution scheme, presented in chapter 3. Since the order parameter, describing the crack, is taken as constant during the solution of the equation for balance of Li, it is not required to introduce new off-diagonal terms in the stiffness matrix. Hence, for these modifications, only comparatively small adjustments in the process of deriving the weak form, performing the linearization, and so on, is required.

for incorporation of additional effects for Li diffusion at the crack. We therefore discuss a small number of possible extensions in the following.

Blockage of Lithium Diffusion

An arbitrarily located crack in the interior of the storage particle, which is not filled with electrolyte, may hamper Li diffusion perpendicular to the crack faces in a significant way. A straight forward implementation of this behavior could be achieved by multiplying the Li flux

$$\mathbf{J} = -D_0 \nabla c + \frac{D_0 \Omega}{R\theta} c\,(1 - c/c_{\max}) \nabla\,\sigma, \tag{2.20}$$

derived in chapter 2, by a prefactor that depends on the local value of the order parameter d, e.g. by

$$\tilde{\mathbf{J}} = [(1-d)^\alpha + d^\alpha k_{\mathrm{res}}]\mathbf{J}, \tag{9.1}$$

with $\alpha > 0$ and $k_{\mathrm{res}} \in [0, 1)$. As a consequence of this modification, the Li flux is decreased at locations of non-zero order parameter, with the relative reduction being determined by k_{res}. This approach is equivalent to a local decrease of the diffusion coefficient in the sense that $\tilde{D} = [(1-d)^\alpha + d^\alpha k_{\mathrm{res}}]D_0$. The assumption may improve the physical correctness of the description of Li diffusion across the crack. However, it does not take into account that Li flux parallel to the crack faces should not be hindered by the crack. In general, similar to grain boundary diffusion in polycrystalline materials, which can dominate over lattice diffusion at low temperatures [183], Li diffusion along the crack faces can be different compared to the one in the bulk material. Such an effect can be approximated by usage of the gradient of the phase field. Except for the location of the crack tip and at possible branching points, it is orientated in perpendicular direction to the crack faces. The modification of the flux thus reads as

$$\tilde{\mathbf{J}} = (1-d)^\alpha \mathbf{J} + d^\alpha (k_1 \mathbf{D} + k_2(\mathbf{1} - \mathbf{D}))\mathbf{J}, \tag{9.2}$$

where $\mathbf{D} = (\nabla d \otimes \nabla d)/d^2$ is a tensor projecting onto the direction of the phase field gradient. Hence, at the location of the crack, the effective diffusion coefficient can be written as $\tilde{D}_1 = k_1 d^\alpha D_0$ for Li flux in perpendicular direction to the crack faces and $\tilde{D}_2 = k_2\, d^\alpha D_0$ for Li flux parallel to the crack faces.

Additional Flux at the Crack Face

In a similar sense as for the blockage of Li diffusion, the order parameter d can be exploited for modeling Li diffusion in the crack by using its value to switch between terms describing Li diffusion in the storage material and terms related to Li transport in the electrolyte within the crack. However, Li flux in the crack depends on several factors that require more thorough considerations. On the one hand, the flow of electrolyte, serving as a medium for Li transport in the crack, is influenced by, for example, its viscosity and capillary effects related to the crack opening. On the other hand, the diffusion of Li in the electrolyte may be modified by the strong spatial confinement in direction perpendicular to the crack faces. Further, the intercalation reaction at the surface of the storage particle requires electrons to recombine with Li-ions so that the formed Li can move into the storage particle. The supply of electrons is influenced by the electrical conductivity of the storage material and the possible coating on its surface. On the crack faces this might be different from the one at the perimeter of the storage particle, for example, due to the lack of a coating layer.

However, a first approximation can be made by assuming that Li transport into the crack and electron conduction along the crack are so fast that they do not limit the insertion (extraction) of Li at the crack faces. In this case, an intercalation (deintercalation) rate of approximately equal magnitude as that on the perimeter of the storage particle can be expected. In this scenario, Li insertion (extraction) at the crack faces could be modeled through placement of the initial crack in a symmetry plane of the storage particle and subsequent simulation of one of the symmetrically equivalent halves of the storage particle. The boundary condition at the symmetry plane would then be defined via

$$\hat{J}(c,d) = d^{\alpha}\bar{J}(c), \tag{9.3}$$

where $\bar{J}(c)$ is the Li flux prescribed at the perimeter of the storage particle. In this case, Li would be inserted (extracted) at the crack faces also during growth of the crack. However, as soon as the crack begins to propagate out of the symmetry plane, for example, due to crack branching, introduction of Li flux through the crack faces would not be applicable with this method since the crack would no longer lie at the boundary. A more versatile approach is therefore presented in the following.

In the formulation of the phase field model for fracture, described in section 2.2, the method for the computation of the fracture energy in the phase field approach was explained. In the context of equation (2.23), we discussed that the actual crack surface is approximated by a volume integration of the crack surface density $\int_{\Gamma(t)} \mathrm{d}A \approx \int_{\mathcal{B}} \mathrm{d}V\, \gamma_l(d)$. This crack surface density can be exploited to describe an additional Li flux through the crack faces during the growth of cracks along arbitrary paths. In the first introduction of the conservation of Li,

$$\frac{\mathrm{d}}{\mathrm{d}t}\int\limits_{\mathcal{P}} \mathrm{d}V\, c(\mathbf{x},t) = -\int\limits_{\partial\mathcal{P}} \mathrm{d}S\, \mathbf{J}(\mathbf{x},t)\cdot\mathbf{n}(\mathbf{x},t), \tag{2.1}$$

given in section 2.1.1, possible internal sources or sinks for Li within the storage particle were neglected. However, if a crack, that permits a constant flux $\bar{J}_0$ into or out of the storage particle, is located at $\mathbf{x} \in \Gamma(t) \subset \mathcal{P}$, an additional term is added to the balance equation so that

$$\frac{\mathrm{d}}{\mathrm{d}t}\int\limits_{\mathcal{P}} \mathrm{d}V\, c(\mathbf{x},t) = 2\bar{J}_0 \int_{\Gamma(t)} \mathrm{d}A - \int\limits_{\partial\mathcal{P}} \mathrm{d}S\, \mathbf{J}(\mathbf{x},t)\cdot\mathbf{n}(\mathbf{x},t), \tag{9.4}$$

where the factor 2 comes from the two faces of a crack. As for the fracture energy, we approximate the crack surface by the integral of the crack surface density. In addition, we write the Li flux $\bar{J}_0$ into the integral to obtain

$$\frac{\mathrm{d}}{\mathrm{d}t}\int\limits_{\mathcal{P}} \mathrm{d}V\, c(\mathbf{x},t) = \int\limits_{\mathcal{P}} \mathrm{d}V\, 2\bar{J}_0\gamma_l(d) - \int\limits_{\partial\mathcal{P}} \mathrm{d}S\, \mathbf{J}(\mathbf{x},t)\cdot\mathbf{n}(\mathbf{x},t). \tag{9.5}$$

The so modified equation for the balance of Li can again be stated in differential form as

$$\dot{c}(\mathbf{x},t) + \nabla\cdot\mathbf{J}(\mathbf{x},t) = 2\bar{J}_0\gamma_l(d), \tag{9.6}$$

where the crack surface density now acts as a source or sink for Li, depending on the sign of $\bar{J}_0$. The method of introducing an additional flux through a suitable boundary condition, as given in equation (9.3), is, to some extent, a special case of this approach. In order to make this statement clearer, we integrate equation (9.3) along the boundary $\mathcal{S} \subset \partial\mathcal{B}_{\mathrm{reduced}}$, which would coincide with the symmetry plane in the full geometry. Here, $\mathcal{B}_{\text{reduced}}$ is the domain describing the reduced geometry, i.e. the symmetric half of the storage particle, and $\partial\mathcal{B}_{\text{reduced}}$ is the corresponding boundary. In addition, we assume that the Li concentration is constant along the boundary. The total Li flux prescribed through $\mathcal{S}$ is then given by

$$\mathrm{J}_{\mathcal{S}}^{\text{total}} = \bar{J}_0 \int_{\mathcal{S}} \mathrm{d}A\, d^{\alpha}. \tag{9.7}$$

For a vanishing length scale parameter l, the spatial transition from $d = 0$ to $d = 1$ becomes negligible and the value of the integral on the right hand side of equation (9.7) converges to the actual value of the total crack surface, i.e.

$$\lim_{l \to 0} \int_{\mathcal{S}} \mathrm{d}A\, d^{\alpha} = \int_{\Gamma(t)} \mathrm{d}A. \tag{9.8}$$

Within the restrictions of the first method, i.e. coincidence of the position of the crack with a symmetry plane of the storage particle, both formulations are therefore identical in an asymptotic sense ($l \to 0$).

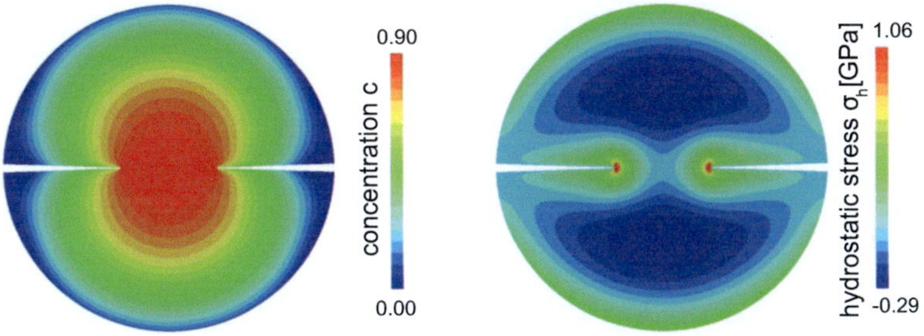

Figure 9.1.: (Left) Concentration c in units of maximum concentration $c_{\max}$ at intermediate time step during Li extraction at the perimeter of the storage particle and the crack faces with $r = 5\,\mu$m, $a = 0.1\,\mu$m and $C = 5$. (Right) The hydrostatic stress σ_h at the same time step. Areas with $d > 0.95$ are removed to show the location of the crack. The opening of the crack is exaggerated by a linear scaling of the nodal displacements.

The modified equation for balance of Li is implemented in the finite elements code and a small number of simulations for Li extraction are performed to show possible consequences of the additional flux at the crack faces. The factor $\bar{J}_0$ is defined consistent to the flux applied at the perimeter of the storage particle, as, for example, described on page 96 in chapter 6. However, as mentioned on page 23 of chapter 2, the computation of the crack surface density $\gamma_l(d)$ in our simulations yields values that are slightly too large. In order to not overestimate the influence of the additional flux at the crack faces, we therefore only apply half of $\bar{J}(c)$ at the crack faces.

The concentration distribution and hydrostatic stress at an intermediate step of crack growth during Li extraction in a storage particle with **r=5 μm, a=0.1 μm** and **C=5** is shown in FIG. 9.1. The effect of the additional extraction is demonstrated by the lower concentration at the crack faces close to the perimeter of the storage particle (compare, for example, the straight isolines of the Li concentration at the cracks in FIG. 7.6 and 7.10 of chapter 7 with the curved ones in FIG. 9.1).

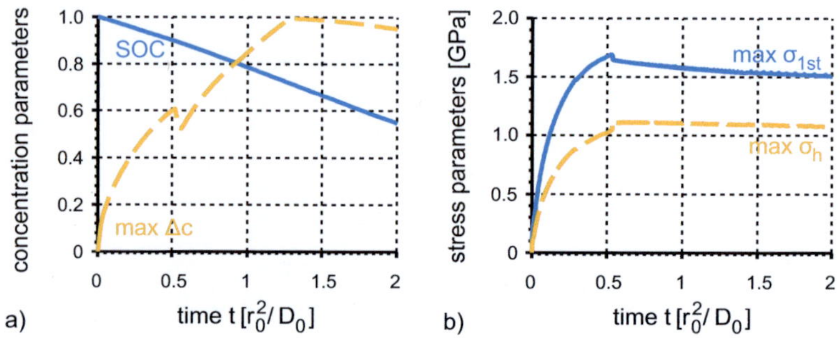

Figure 9.2.: Characteristic results during Li extraction at the perimeter of storage particle and the crack faces with $r = 5\,\mu$m, $a = 0.1\,\mu$m and $C = 5$. (a) State of charge (SOC) and maximum concentration difference $\max \Delta c$ in units of $c_{\max}$ in units of maximum concentration $c_{\max}$. (b) The highest value of the maximum principal stress $\max \sigma_{1\text{st}}$ and maximum hydrostatic stress $\max \sigma_h$ during Li extraction.

The impact of the additional Li extraction is also visible in the time history of the maximum concentration difference $\max \Delta c$, plotted in FIG. 9.2 a). The corresponding graph first rises until the crack begins to grow in an initially unstable way at $t \approx 0.5\,\tau$. The unstable crack propagation is further marked by the peak in the highest value of the maximum principal stress $\max \sigma_{1\text{st}}$ in 9.2 b). Due to the short time scale of unstable crack growth and the fast change in the hydrostatic stress field, the maximum concentration difference then shortly drops before it grows again. The steeper slope of the graph of the maximum concentration difference, observed after crack growth, stems from the increased area through which Li is extracted. This leads to a more rapid drop of the Li concentration, in particular close to the location at the perimeter of the storage particle where the initial cracks are introduced. The higher total rate at which Li is extracted also leads to higher stress in the particle. This is reflected by the larger final crack extension during Li extraction with additional flux through the crack faces ($\Delta a \approx 3.50\,\mu\text{m}$) in comparison to that without additional flux ($\Delta a \approx 3.23\,\mu\text{m}$).

Increasing the C-rate to $\mathbf{C = 10}$, the extraction of Li at the crack faces leads to more drastic consequences on the final crack geometry. As shown in FIG. 9.3, the additional Li extraction generates a significant gradient in the Li concentration close to the crack tip. This gradient generates sufficient mechanical stress to induce crack branching. The new branches then propagate towards the perimeter of the storage particle in almost perfectly perpendicular orientation with respect to the prior direction of crack growth. In contrast to the study on crack growth during Li insertion in chapter 6, this branching is not caused by inertia effects but results solely from the distribution of mechanical stress induced by the inhomogeneous Li concentration.

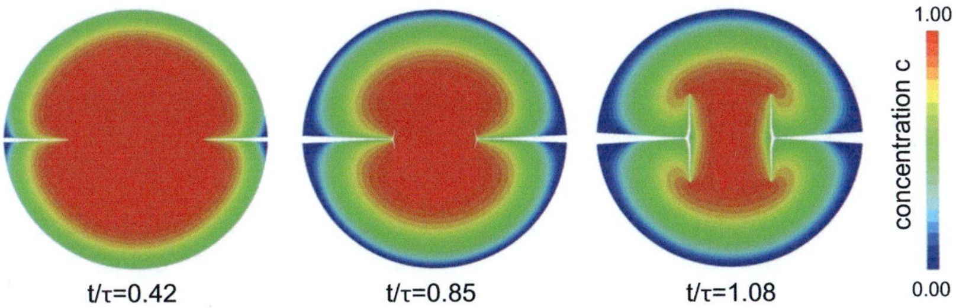

Figure 9.3.: Concentration c in units of maximum concentration c_{max} for different times during Li extraction at the perimeter of the storage particle and the crack faces with $r = 5\,\mu\text{m}$, $a = 0.1\,\mu\text{m}$ and $C = 10$. Times are given in units of $\tau = r_0^2/D_0$. Areas with $d > 0.95$ are removed to show the location of the crack. In addition, the geometry is scaled using the nodal displacement values.

Due to the deep penetration length of the cracks into the storage particle and because of the crack branching in its interior, the particle may become too fragile to withstand the stress generated in the next half cycle during Li insertion and break. However, we observe that this in not the case if the boundary conditions are applied in the same way during Li insertion in the second half cycle. This is shown by contour plots of the resulting Li concentration in FIG. 9.4. Although the total rate of Li insertion is faster due to the additional Li insertion at the crack faces, the difference of the Li concentration between outer and inner parts of the particle is smaller because Li is also inserted close to the center of the storage particle. As a consequence, the mechanical stress resulting from the inhomogeneous Li concentration distribution does not acquire sufficiently large magnitudes to induce further crack growth.

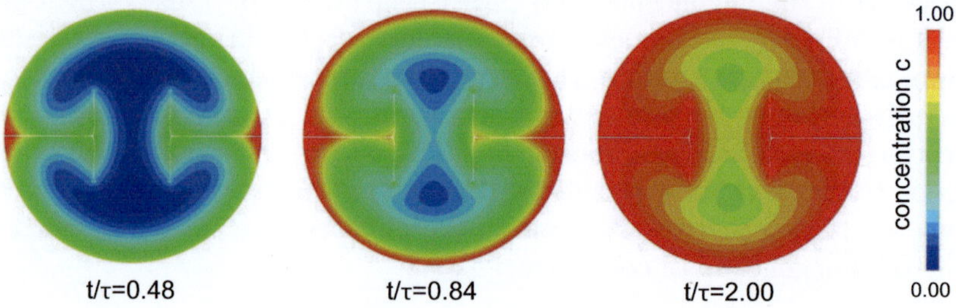

Figure 9.4.: Concentration c in units of maximum concentration c_{max} during Li insertion in the second half cycle of the storage particle shown in FIG. 9.3. Times are given in units of $\tau = r_0^2/D_0$. Areas with $d > 0.95$ are removed to show the location of the crack. In addition, the geometry is scaled using the nodal displacement values.

Comparison of FIG. 9.3 and 9.4 also gives an indication of the crack opening due to the different stress fields during extraction and insertion of Li. In FIG. 9.3, the cracks are under tensile stress, which is reflected by the larger crack opening. In contrast, during Li insertion, the mechanical stress is mainly compressive at the location of the cracks. Hence, the cracks do not open. As a result, it is doubtful whether the cracks are filled by electrolyte so that Li can be inserted through the crack faces. A more elaborate description of Li diffusion into and along the crack becomes necessary. This is not pursued in this work.

10 Conclusion

A coupled model of Li diffusion, mechanical stress and crack growth is presented using a phase field method for fracture. The approach allows the study of the behavior of Li diffusion and crack growth, as well as their mutual interaction, in a simultaneous fashion. Due to the numerical complexity of the problem, the system is implemented in an elaborate computer algorithm.

The solution approach, as well as implementation details, are presented and the phase field method for crack growth is compared with relations from fracture mechanics. We find that results of the phase field simulations are in very good agreement if the involved length scale parameter is small and the initial crack is induced as a heterogeneity in the phase field through an initially non-zero history field. Under these conditions the phase field method is assessed as a suitable approach to study crack growth coupled to the diffusion of Li in storage particles. We therefore apply the model to investigate crack growth during Li insertion and extraction, as well as in the second half cycle with opposite direction of the applied flux.

Crack growth starting at the interior of a storage particle during insertion of Li is highly influenced by inertia effects due to the strongly unstable manner of crack propagation. Although material properties are assumed to provide a setting that is as simple as possible, i.e. without taking into account, for example, crystallographic planes, non-isotropic diffusion, phase changes, and much more, the interplay of Li diffusion and fracture, as well as the resulting crack topologies, become rather complex.

The comparison between two and three dimensional computations reveals a number of similarities, e.g. the fundamental dependencies on particle and crack size and the behavior of characteristic stresses and concentrations. At the same time, results of 3D simulations become more intricate due to the additional direction possible for high components of stress. We observe, that although one crack may run through almost the whole particle, it might not relieve sufficient stress in other directions to avoid additional cracks from growing.

We find, that growth of an initial crack in the interior of the particle can lead to breakage of the particle in only one insertion half cycle. Depending on the nature of crack branching, the particle may thereby be fragmented into several parts. These effects are most pronounced for large particles, high C-rates and small crack sizes.

When the initial precrack is comparatively large, the crack does not penetrate to the perimeter and a short regime of stable crack growth typically follows the initially unstable extension. However, the thin, remaining ligaments, joining the different parts, are believed to fail in a subsequent extraction half cycle. This is confirmed by exemplary simulations of the subsequent extraction half cycle.

Growth of an initial surface crack during Li extraction is found to occur under less severe conditions than during insertion, i.e. in smaller storage particles or with slower Li extraction. For the most part, cracks grow in a stable way towards the particle interior and come to rest with their tip at some distance from the center. Depending on the C-rate, particle radius and initial crack size, a phase of unstable crack growth may precede the stable regime.

This remains true in a three-dimensional spherical particle. However, in comparison to the two-dimensional geometry, the crack first grows in an unstable fashion along the surface and then penetrates in a ring-like shape into the particle. The symmetry and geometry of the particle are found to be more decisive on the final crack shape than the initial crack form.

Breakage of the particle during a single extraction half cycle is not observed. However, in most of the simulations showing crack growth, the final location of the crack tip is situated at a radius that would experience tensile stress in a subsequent half cycle of Li insertion. We confirm the occurrence of crack growth and storage particle fragmentation in simulations of the second half cycle.

From this we conclude that particle breakage both occurs in a single step, i.e. within one half cycle, or in a process of two or many steps, even without taking into account additional fatigue mechanisms that might pile up during cycling. These findings are expected to hold in a similar way for more elaborate material descriptions such as, for example, those taking into account phase transformations.

A Appendix

A.1 Non-Dimensionalization

The coupled system of partial differential equations, shown in FIG. 2.2 requires a set of material parameters. These parameters determine the ratio in magnitude of different physical effects, e.g. the ratio between chemically induced and mechanically driven Li flux. However, as explained why the Buckingham π-theorem in chapter 6, they do not describe a fully independent set, in the sense that the described physical problem only depends on a subclass of combinations of the parameters. To some extent the parameters therefore only scale the corresponding equations. For example, the conservation of linear momentum in a static situation, i.e. $\nabla \cdot \sigma = 0$, is independent of Young's modulus E if the boundary conditions are of Dirichlet type, i.e. with prescribed displacements. As a matter of fact, the resulting stress in the material is not independent from E. However, in order to find a solution for the displacement field $\mathbf{u}$, the value of E is not required. Indeed, in a numerical implementation, it is generally recommended to avoid scaling of the equations by material parameters since they can lead to larger numerical errors, when prefactors are very small or large. Further, numerical parameters, e.g. absolute convergence thresholds, have to be adjusted if the corresponding equations scale with a factor.

For this reason, we implement the boundary value problem in a non-dimensionalized form. As mentioned in chapter 6, there is some degree of freedom in the choice of the non-dimensionalized parameters. In contrast to [167], our choice is not based on physical arguments, but on simple substitution rules for a straight-forward numerical implementation.

We begin with the balance of Li given in (2.3). The Li concentration c ranges from 0 to $c_{\max}$ in our model. To keep numbers close to 1, it is therefore reasonable to express c in terms of $c_{\max}$. The spatial gradient and the time derivative introduce two dimensions that are scaled by the characteristic length r_0, e.g. the radius of a particle, and time $\tau = r_0^2/D_0$. The substitution rules thus read

$$\begin{aligned}
c &\mapsto \hat{c} = \frac{c}{c_{\max}}, & t &\mapsto \hat{t} = t\frac{D_0}{r_0^2}, & \partial_t &\mapsto \partial_{\hat{t}} = \frac{\partial t}{\partial \hat{t}}\partial_t = \frac{r_0^2}{D_0}\partial_t, \\
\mathbf{J} &\mapsto \hat{\mathbf{J}} = \mathbf{J}\frac{r_0}{D_0 c_{\max}}, & \mathbf{x} &\mapsto \hat{\mathbf{x}} = \frac{\mathbf{x}}{r_0} & \nabla_{\mathbf{x}} &\mapsto \nabla_{\hat{\mathbf{x}}} = \frac{\partial \mathbf{x}}{\partial \hat{\mathbf{x}}}\nabla_{\mathbf{x}} = r_0 \nabla_{\mathbf{x}},
\end{aligned} \tag{A.1}$$

where the replacement of the flux $\mathbf{J}$ follows for reasons of consistency. The balance of Li then reads

$$\partial_{\hat{t}}\hat{c} + \nabla_{\hat{\mathbf{x}}}\hat{\mathbf{J}} = 0. \tag{A.2}$$

Boundary conditions, such as an applied flux at the boundary must be given in the same form, i.e.

$$\bar{\mathbf{J}} \mapsto \hat{\bar{\mathbf{J}}} = \bar{\mathbf{J}}\, r_0/(D_0 c_{\max}). \tag{A.3}$$

In contrast to the material parameters, which are taken from the literature for $LiMn_2O_4$, the characteristic length is fixed to $r_0 = 1\,\mu\text{m}$.

Applying the rules to the definition of the flux (2.20) leaves a last free choice on how to substitute the material parameters. We wish to keep the non-dimensionality of the product describing volume changes due to Li, Ωc, and define the molar volume Ω in reciprocal units of $c_{\max}$. Again, to maintain consistency, this yields a set of new replacement rules

$$\begin{aligned}
\Omega &\mapsto \hat{\Omega} = \Omega\, c_{\max}, & \boldsymbol{\sigma} &\mapsto \hat{\boldsymbol{\sigma}} = \frac{\boldsymbol{\sigma}}{R\theta/\Omega}, & L &\mapsto \hat{L} = \frac{L}{R\theta/\Omega}, \\
G' &\mapsto \hat{G}' = \frac{G'}{R\theta/\Omega}, & \nu &\mapsto \hat{\nu} = \nu, & E &\mapsto \hat{E} = \frac{E}{R\theta/\Omega}.
\end{aligned} \tag{A.4}$$

To avoid confusion, we emphasize that G' is the shear modulus and not the energy release rate. The same substitution rule for the stress and the elastic parameters results from the non-dimensionality of the strain $\boldsymbol{\varepsilon}$

$$\begin{aligned}
&\mathbf{u} \mapsto \hat{\mathbf{u}} = \mathbf{u}/r_0, \\
&\boldsymbol{\varepsilon} \mapsto \hat{\boldsymbol{\varepsilon}} = \text{sym}[\nabla_{\hat{\mathbf{x}}}\hat{\mathbf{u}}] = \text{sym}[r_0\nabla_{\mathbf{x}}\mathbf{u}/r_0] = \text{sym}[\nabla_{\mathbf{x}}\mathbf{u}] = \boldsymbol{\varepsilon}.
\end{aligned} \tag{A.5}$$

The non-dimensionalized flux then reads

$$\hat{\mathbf{J}} = -\nabla_{\hat{\mathbf{x}}}\hat{c} + \hat{c}(1-\hat{c})\nabla_{\hat{\mathbf{x}}}\hat{\sigma}_h. \tag{A.6}$$

In the same way, we proceed with the conservation of linear momentum in equation (2.7) to receive

$$\hat{\rho}\,\partial_{\hat{t}}\hat{\mathbf{v}} = \nabla_{\hat{\mathbf{x}}}\cdot\hat{\boldsymbol{\sigma}}, \tag{A.7}$$

with the substituted density and velocity given by

$$\rho \mapsto \hat{\rho} = \rho\frac{D_0^2}{r_0^2}\frac{\Omega}{R\theta}, \quad \mathbf{v} \mapsto \hat{\mathbf{v}} = \frac{r_0}{D_0}\mathbf{v}, \quad \dot{\mathbf{v}} \mapsto \partial_{\hat{t}}\hat{\mathbf{v}} = \frac{r_0^3}{D_0^2}\dot{\mathbf{v}}. \tag{A.8}$$

The replacement rules for the elastic parameters apply in the same way for all kinds of energies, so that with

$$\begin{aligned} \psi \mapsto \hat{\psi} &= \frac{\psi}{R\theta/\Omega}, & \mathcal{H} \mapsto \hat{\mathcal{H}} &= \frac{\mathcal{H}}{R\theta/\Omega}, \\ l \mapsto \ell &= l/r_0, & G_c \mapsto \hat{G}_c &= \frac{G_c}{r_0 R\theta/\Omega}, \end{aligned} \tag{A.9}$$

the evolution equation of the phase field (2.29) takes the form

$$\frac{\hat{G}_c}{\ell}\left[d - \ell^2\nabla_{\hat{\mathbf{x}}}^2 d\right] = 2(1-d)\hat{\mathcal{H}}. \tag{A.10}$$

A.2 Exact Decomposition of Elastic Energy Density

In order to demonstrate that the decomposition of the elastic energy, introduced in equation (2.27), is exact in the sense that, with vanishing order parameter, i.e. $d = 0$, and neglecting the numerical parameter k, the sum of tensile and compressive parts yields the ordinary tensile energy, as defined by

$$\psi(\boldsymbol{\varepsilon}) = \frac{1}{2}L(\mathrm{tr}[\boldsymbol{\varepsilon}])^2 + G'\mathrm{tr}[\boldsymbol{\varepsilon}^2], \tag{A.11}$$

we add up the compressive and tensile part of the decomposed energy by

$$\psi^+(\varepsilon) + \psi^-(\varepsilon) = \frac{1}{2}L\langle\mathrm{tr}[\varepsilon]\rangle_+^2 + G'\mathrm{tr}[\langle\varepsilon\rangle_+^2] + \frac{1}{2}L\langle\mathrm{tr}[\varepsilon]\rangle_-^2 + G'\mathrm{tr}[\langle\varepsilon\rangle_-^2]. \tag{A.12}$$

Next, we rearrange the terms to receive

$$\psi^+(\varepsilon) + \psi^-(\varepsilon) = \frac{1}{2}L\left(\langle\mathrm{tr}[\varepsilon]\rangle_+^2 + \langle\mathrm{tr}[\varepsilon]\rangle_-^2\right) + G'\left(\mathrm{tr}[\langle\varepsilon\rangle_+^2] + \mathrm{tr}[\langle\varepsilon\rangle_-^2]\right). \tag{A.13}$$

For the volumetric parts, we express the trace operator in terms of the eigenvalues, which yields

$$\langle\mathrm{tr}[\varepsilon]\rangle_+^2 + \langle\mathrm{tr}[\varepsilon]\rangle_-^2 = \langle\sum_i \lambda_i\rangle_+^2 + \langle\sum_i \lambda_i\rangle_-^2 = (\sum_i \lambda_i)^2 = (\mathrm{tr}[\varepsilon])^2. \tag{A.14}$$

We proceed in the same way for the deviatoric part

$$\begin{aligned}
\mathrm{tr}[\langle\varepsilon\rangle_+^2] + \mathrm{tr}[\langle\varepsilon\rangle_-^2] &= \mathrm{tr}[\langle\sum_i \lambda_i\boldsymbol{\eta}_i\rangle_+^2] + \mathrm{tr}[\langle\sum_i \lambda_i\boldsymbol{\eta}_i\rangle_-^2] \\
&= \mathrm{tr}[(\sum_i \langle\lambda_i\rangle_+\boldsymbol{\eta}_i)^2] + \mathrm{tr}[(\sum_i \langle\lambda_i\rangle_-\boldsymbol{\eta}_i)^2] \\
&= \mathrm{tr}[(\sum_i \langle\lambda_i\rangle_+^2\boldsymbol{\eta}_i)] + \mathrm{tr}[(\sum_i \langle\lambda_i\rangle_-^2\boldsymbol{\eta}_i)] \\
&= \sum_i \langle\lambda_i\rangle_+^2 + \sum_i \langle\lambda_i\rangle_-^2 \\
&= \sum_i \lambda_i^2,
\end{aligned} \tag{A.15}$$

where we use the representation of ε in terms of its eigenvectors and eigenprojectors and the definition of the bracket operators, as well as the one for powers of a matrix. We express the result again in a different way by

$$\sum_i \lambda_i^2 = \mathrm{tr}[(\sum_i \lambda_i^2\boldsymbol{\eta}_i)] = \mathrm{tr}[\varepsilon^2]. \tag{A.16}$$

Inserting equations (A.14-A.16) into equation (A.13) then yields

$$\psi^+(\varepsilon) + \psi^-(\varepsilon) = \frac{1}{2}L(\mathrm{tr}[\varepsilon])^2 + G'\mathrm{tr}[\varepsilon^2] = \psi(\varepsilon). \tag{A.17}$$

A.3 Cauchy Stress Tensor for Spectrally Decomposed Strain

Defined as the derivative of the free energy with respect to the full strain, we calculate the stress tensor as

$$\boldsymbol{\sigma} = \frac{\partial}{\partial \boldsymbol{\varepsilon}} \psi_{\text{mech}}(\boldsymbol{\varepsilon}, c) = [g(d) + k] \frac{\partial}{\partial \boldsymbol{\varepsilon}} \psi^{+}_{\text{intact}}(\boldsymbol{\varepsilon}, c) + \frac{\partial}{\partial \boldsymbol{\varepsilon}} \psi^{-}_{\text{intact}}(\boldsymbol{\varepsilon}, c), \quad \text{(A.18)}$$

with

$$\frac{\partial}{\partial \boldsymbol{\varepsilon}} \psi^{\pm}_{\text{intact}}(\boldsymbol{\varepsilon}, c) = L \langle \text{tr}[\boldsymbol{\varepsilon} - \boldsymbol{\varepsilon}_{\text{Li}}] \rangle_{\pm} \mathbf{1} + 2G' \langle \boldsymbol{\varepsilon} - \boldsymbol{\varepsilon}_{\text{Li}} \rangle_{\pm}. \quad \text{(A.19)}$$

For the first term, we used

$$\frac{\partial}{\partial \boldsymbol{\varepsilon}} \text{tr}^2[\boldsymbol{\varepsilon} - \boldsymbol{\varepsilon}_{\text{Li}}] = \frac{\partial \text{tr}^2[\boldsymbol{\varepsilon} - \boldsymbol{\varepsilon}_{\text{Li}}]}{\partial \text{tr}[\boldsymbol{\varepsilon} - \boldsymbol{\varepsilon}_{\text{Li}}]} \frac{\partial \text{tr}[\boldsymbol{\varepsilon} - \boldsymbol{\varepsilon}_{\text{Li}}]}{\partial \boldsymbol{\varepsilon}} = 2\text{tr}[\boldsymbol{\varepsilon} - \boldsymbol{\varepsilon}_{\text{Li}}] \mathbf{1}, \quad \text{(A.20)}$$

and for the second term

$$\begin{aligned} \frac{\partial}{\partial \boldsymbol{\varepsilon}} \text{tr}[\langle \boldsymbol{\varepsilon} - \boldsymbol{\varepsilon}_{\text{Li}} \rangle^2] &= \frac{\partial}{\partial \boldsymbol{\varepsilon}} \sum_i \langle \tilde{\lambda}_i \rangle^2_{\pm} \\ &= 2 \sum_i \langle \tilde{\lambda}_i \rangle_{\pm} \frac{\partial}{\partial \boldsymbol{\varepsilon}} \langle \tilde{\lambda}_i \rangle_{\pm} \\ &= 2 \sum_i \langle \tilde{\lambda}_i \rangle_{\pm} \tilde{\boldsymbol{\eta}}^{(i)} \\ &= 2 \langle \boldsymbol{\varepsilon} - \boldsymbol{\varepsilon}_{\text{Li}} \rangle_{\pm}. \end{aligned} \quad \text{(A.21)}$$

In the second last step, we have exploited a relationship between the derivative of an eigenvalue with respect to the related tensor and the corresponding eigenspace projection operator, see Lemma 2 below.

Lemma 1

If Φ is isotropic, then

$$\frac{\partial \Phi}{\partial \boldsymbol{\varepsilon}} = \sum_i \frac{\partial \Phi}{\partial \lambda_i} \boldsymbol{\eta}^{(i)}.$$

Prove: If Φ is isotropic, one can use the three invariants

$$\alpha_j = j^{-1}\mathrm{tr}\boldsymbol{\varepsilon}^j, \qquad \text{with} \qquad j = 1, 2, 3$$

to express the dependence of Φ on the strain tensor by

$$\Phi(\boldsymbol{\varepsilon}) = \Phi(\alpha_1, \alpha_2, \alpha_3).$$

For the derivative of Φ we then use the chain rule

$$\frac{\partial \Phi}{\partial \boldsymbol{\varepsilon}} = \sum_j \frac{\partial \Phi}{\partial \alpha_j}\frac{\partial \alpha_j}{\partial \boldsymbol{\varepsilon}} = \sum_j \frac{\partial \Phi}{\partial \alpha_j}\boldsymbol{\varepsilon}^{j-1}.$$

The last step becomes clear by applying j times $\frac{\partial}{\partial \mathbf{A}}\mathrm{tr}[\mathbf{AB}] = \mathbf{B}^t$ or $\frac{\partial}{\partial a_{ij}}\mathrm{tr}[\mathbf{AB}] = \frac{\partial}{\partial a_{ij}}\sum_{lm} a_{lm}b_{ml} = b_{ji}$ with $\mathbf{A} = \boldsymbol{\varepsilon}$ and $\mathbf{B} = \mathbf{B}^t = \boldsymbol{\varepsilon}^{j-1}$. To close the prove, we use the spectral decomposition

$$\boldsymbol{\varepsilon}^{j-1} = \sum_i \lambda_i^{j-1}\boldsymbol{\eta}^{(i)},$$

as well as the dependance of the three invariants on the eigenvalues

$$\alpha_j = j^{-1}\sum_i \lambda_i^j,$$

which gives us the derivatives

$$\frac{\partial \alpha_j}{\partial \lambda_i} = \lambda_i^{j-1}.$$

Putting everything together, we receive

$$\begin{aligned}
\frac{\partial \Phi}{\partial \boldsymbol{\varepsilon}} &= \sum_j \frac{\partial \Phi}{\partial \alpha_j}\boldsymbol{\varepsilon}^{j-1} \\
&= \sum_i \sum_j \frac{\partial \Phi}{\partial \alpha_j}\lambda_i^{j-1}\boldsymbol{\eta}^{(i)} \\
&= \sum_i \sum_j \frac{\partial \Phi}{\partial \alpha_j}\frac{\partial \alpha_j}{\partial \lambda_i}\boldsymbol{\eta}^{(i)} \\
&= \sum_i \frac{\partial \Phi}{\partial \lambda_i}\boldsymbol{\eta}^{(i)}.
\end{aligned}$$

Lemma 2

If λ_i is an eigenvalue of $\boldsymbol{\varepsilon}$, then

$$\frac{\partial \lambda_i}{\partial \boldsymbol{\varepsilon}} = \boldsymbol{\eta}^{(i)}.$$

Since an eigenvalue is an objective quantity, it can be expressed by the three invariants of *Lemma 1*. This can be shown explicitly in a few steps. Hence, λ_i fulfills the required properties of Φ in *Lemma 1*. We therefore set $\Phi = \lambda_i$ and confirm the above statement by using $\frac{\partial \lambda_i}{\partial \lambda_j} = \delta_{ij}$.

Bibliography

[1] R. A. Huggins. *Advanced batteries.* Springer Verlag, 2009.

[2] C. Daniel and J. O. Besenhard. *Handbook of battery materials.* Wiley-VCH, 2011.

[3] CTIA's Semi-Annual Wireless Industry Survey. `http://www.ctia.org/`. Accessed: 2014-01-22.

[4] Avicenne. The Portable Rechargeable Battery (PRB) market in Europe (2008-2015) (presented at BatteriesEvent2010). `http://www.rechargebatteries.org/`. Accessed: 2013-11-12.

[5] Electric Bicycles - Global Market Opportunities, Barriers, Technology Issues, and Demand Forecasts for E-Bicycles, Pedal-Assist Bicycles, and E-Bicycle Batteries and Motors. `http://www.navigantresearch.com/`. Accessed: 2014-01-23.

[6] Zahlen - Daten - Fakten zum Fahrradmarkt in Deutschland. `http://www.ziv-zweirad.de/`. Accessed: 2014-01-22.

[7] Roland Berger. Technology & Market Drivers for Stationary and Automotive Battery Systems (presented at BatteriesEvent2012). `http://www.rechargebatteries.org/`. Accessed: 2013-11-12.

[8] The development of e-mobility in global markets (presented at IAA2013). `http://archiv.iaa.de/`. Accessed: 2014-01-31.

[9] Battery technology charges ahead. `http://www.mckinsey.com/`. Accessed: 2014-02-03.

[10] Technical facts VW e-up! `http://www.volkswagen.de/`. Accessed: 2014-02-03.

[11] Technical facts BMW i3. `http://www.bmw.de/`. Accessed: 2014-02-04.

[12] L. Lu, X. Han, J. Li, J. Hua, and M. Ouyang. A review on the key issues for lithium-ion battery management in electric vehicles. *Journal of Power Sources*, 226:272–288, 2013.

[13] H. Wallentowitz and A. Freialdenhoven. *Strategien zur Elektrifizierung des Antriebsstranges.* Vieweg+Teubner Verlag, 2011.

[14] M. S. Whittingham. Electrical Energy Storage and Intercalation Chemistry. *Science*, 192:1126–1127, 1976.

[15] J. O. Besenhard and H. P. Fritz. Cathodic reduction of graphite in organic solutions of alkali and NR_4^+ salts. *Journal of Electroanalytical Chemistry and Interfacial Electrochemistry*, 53:329–333, 1974.

[16] J. O. Besenhard and G. Eichinger. High energy density lithium cells: Part I. Electrolytes and anodes. *Journal of Electroanalytical Chemistry and Interfacial Electrochemistry*, 68:1–18, 1976.

[17] G. Eichinger and J. O. Besenhard. High energy density lithium cells: Part II. Cathodes and complete cells. *Journal of Electroanalytical Chemistry and Interfacial Electrochemistry*, 72:1–31, 1976.

[18] A. Yoshino. The Birth of the Lithium-Ion Battery. *Angewandte Chemie International Edition*, 51:5798–5800, 2012.

[19] J. C. Hunter. Preparation of a new crystal form of manganese dioxide: λ-MnO_2. *Journal of Solid State Chemistry*, 39:142–147, 1981.

[20] M. Winter, J. O. Besenhard, M. E. Spahr, and P. Novák. Insertion Electrode Materials for Rechargeable Lithium Batteries. *Advanced Materials*, 10(10):725–763, 1998.

[21] M. S. Whittingham. Lithium Batteries and Cathode Materials. *Chemical Review*, 104:4271–4301, 2004.

[22] J. Xu, S.-L. Chou, H.-K. Liu, and S.-X. Dou. The effect of different binders on electrochemical properties of $LiNi_{1/3}Mn_{1/3}Co_{1/3}O_2$ cathode material in lithium ion batteries. *Journal of Power Sources*, 225:172–178, 2013.

[23] W.-J. Zhang. Lithium insertion/extraction mechanism in alloy anodes for lithium-ion batteries. *Journal of Power Sources*, 196:877–885, 2011.

[24] F. Wang, R. Robert, N. A. Chernova, N. Pereira, F. Omenya, F. Badway, X. Hua, M. Ruotolo, R. Zhang, L. Wu, V. Volkov, D. Su, B. Key, M. S. Whittingham, C. P. Grey, G. G. Amatucci, Y. Zhu, and J. Graetz. Conversion Reaction Mechanisms in Lithium Ion Batteries: Study of the Binary Metal Fluoride Electrodes. *Journal of the American Chemical Society*, 13(46):8828–36, 2011.

[25] E. Ferg, R. J. Gummow, A. de Kock, and M. M. Thackeray. Spinel Anodes for Lithium-Ion Batteries. *Journal of the Electrochemical Society*, 11(141):147–150, 1994.

[26] X. L. Yao, S. Xie, C. H. Chen, Q. S. Wang, J. H. Sun, Y. L. Li, and S. X. Lu. Comparisons of graphite and spinel $Li_{1.33}Ti_{1.67}O_4$ as anode materials for rechargeable lithium-ion batteries. *Electrochimica Acta*, 50:4076–4081, 2005.

[27] S. B. Chikkannanavar, D. M. Bernardi, and L. Liu. A review of blended cathode materials for use in Li-ion batteries. *Journal of Power Sources*, 248:91–100, 2014.

[28] J. Vetter, P. Novak, M. R. Wagner, C. Veit, K.-C. Möller, J. O. Besenhard, M. Winter, M. Wohlfahrt-Mehrens, C. Vogler, and A. Hammouche. Ageing mechanisms in lithium-ion batteries. *Journal of Power Sources*, 147:269–281, 2005.

[29] Y. Dai, L. Cai, and R. E. White. Capacity Fade Model for Spinel $LiMn_2O_4$ Electrode. *Journal of the Electrochemical Society*, 160(1):A182–A190, 2013.

[30] M. B. Pinson and M. Z. Bazant. Theory of SEI Formation in Rechargeable Batteries: Capacity Fade, Accelerated Aging and Lifetime Prediction. *Journal of the Electrochemical Society*, 160(2):A243–A250, 2013.

[31] S. P. V. Nadimpalli, V. A. Sethuraman, S. Dalavi, B. Lucht, M. J. Chon, V. B. Shenoy, and P. R. Guduru. Quantifying capacity loss

due to solid-electrolyte-interphase layer formation on silicon negative electrodes in lithium-ion batteries. *Journal of Power Sources*, 215:145–151, 2012.

[32] A. M. Colclasure, K. A. Smith, and R. J. Kee. Modeling detailed chemistry and transport for solid-electrolyte-interace (SEI) films in Li-ion batteries. *Electrochimica Acta*, 58:33–43, 2011.

[33] J. Lei, L. Li, R. Kostecki, R. Muller, and F. McLarnon. Characterization of SEI Layers on $LiMn_2O_4$ Cathodes with *In Situ* Spectroscopic Ellipsometry. *Journal of the Electrochemical Society*, 152(4):A774–A777, 2005.

[34] B. Stiaszny, J. C. Ziegler, E. E. Krauß, M. Zhang, J. P. Schmidt, and E. Ivers-Tiffée. Electrochemical characterization and post-mortem analysis of aged $LiMn_2O_4Li(Ni_{0.5}Mn_{0.3}Co_{0.2})O_2$/graphite lithium ion batteries. Part II: Calendar aging. *Journal of Power Sources*, 258:61–75, 2014.

[35] B. Stiaszny, J. C. Ziegler, E. E. Krauß, J. P. Schmidt, and E. Ivers-Tiffée. Electrochemical characterization and post-mortem analysis of aged $LiMn_2O_4Li(Ni_{0.5}Mn_{0.3}Co_{0.2})O_2$/graphite lithium ion batteries. Part I: Cycle aging. *Journal of Power Sources*, 251:439–450, 2014.

[36] D. Aurbach. Review of selected electrode-solution interactions which determine the performance of Li and Li ion batteries. *Journal of Power Sources*, 89:206–218, 2000.

[37] L. Cai, Y. Dai, M. Nicholson, R. E. White, K. Jagannathan, and G. Bhatia. Life modeling of a lithium ion cell with a spinel-based cathode. *Journal of Power Sources*, 221:191–200, 2013.

[38] J. Park, J. H. Seo, G. Plett, W. Lu, and A. M. Sastry. Numerical Simulation of the Effect of the Dissolution of $LiMn_2O_4$ Particles on Li-Ion Battery Performance. *Electrochemical and Solid-State Letters*, 14(2):A14–A18, 2011.

[39] M. Hirayama, H. Ido, K. Kim, W. Cho, K. Tamura, J. Mizuki, and R. Kanno. Dynamic Structural Changes at $LiMn_2O_4$/Electrolyte Interface during Lithium Battery Reaction. *Journal of the American Chemical Society*, 132:15268–15276, 2010.

[40] H. Shin, J. Park, A. M. Sastry, and W. Lu. Degradation of the solid electrolyte interphase induced by the deposition of manganese ions. *Journal of Power Sources*, 284:416–427, 2015.

[41] D. Mohanty, S. Kalnaus, R. A. Meisner, K. J. Rhodes, J. Li, E. A. Payzant, D. I. Wood III, and C. Daniel. Structural transformation of a lithium-rich $Li_{1.2}Co_{0.1}Mn_{0.55}Ni_{0.15}O_2$ cathode during high voltage cycling resolved by in situ X-ray diffraction. *Journal of Power Sources*, 229:239–248, 2013.

[42] S. K. Martha, J. Nanda, G. M. Veith, and N. J. Dudney. Electrochemical and rate performance study of high-voltage lithium-rich composition: $Li_{1.2}Mn_{0.525}Ni_{0.175}Co_{0.1}O_2$. *Journal of Power Sources*, 199:220–226, 2012.

[43] K. G. Gallagher, J. R. Croy, M. Balasubramanian, M. Bettge, D. P. Abraham, A. K. Burrell, and M. M. Thackeray. Correlating hysteresis and voltage fade in lithium- and manganese-rich layered transition-metal oxide electrodes. *Electrochemistry Communications*, 33:96–98, 2013.

[44] M. Hu, X. Pang, and Z. Zhou. Recent progress in high-voltage lithium ion batteries. *Journal of Power Sources*, 237:229–242, 2013.

[45] Z. Li, J. Huang, B. Y. Liaw, V. Metzler, and J. Zhang. A review of lithium deposition in lithium-ion and lithium metal secondary batteries. *Journal of Power Sources*, 254:168–182, 2014.

[46] J. Steiger, D. Kramer, and R. Mönig. Mechanisms of dendritic growth investigated by in situ light microscopy during electrodeposition and dissolution of lithium. *Journal of Power Sources*, 261:112–119, 2014.

[47] S. Tippmann, D. Walper, L. Balboa, B. Spier, and W. G. Bessler. Low-temperature charging of lithium-ion cells part I: Electrochemical modeling and experimental investigation of degradation behavior. *Journal of Power Sources*, 252:305–316, 2014.

[48] J. Remmlinger, S. Tippmann, M. Buchholz, and K. Dietmayer. Low-temperature charging of lithium-ion cells Part II: Model reduction and application. *Journal of Power Sources*, 254:268–276, 2014.

[49] R. Fu, M. Xiao, and S.-Y. Choe. Modeling, validation and analysis of mechanical stress generation changes of a pouch type high power Li-ion battery. *Journal of Power Sources*, 224:211–224, 2013.

[50] H. Zheng, L. Tan, X. Song, and V. S. Battaglia. Calendering effects on the physical and electrochemical properties of $Li[Ni_{1/3}Mn_{1/3}Co_{1/3}]O_2$ cathode. *Journal of Power Sources*, 208:52–57, 2012.

[51] J. Cannarella and C. B. Arnold. Stress evolution and capacity fade in constrained lithium-ion pouch cells. *Journal of Power Sources*, 245:745–751, 2014.

[52] J. Cannarella and C. B. Arnold. State of health and charge measurements in lithium-ion batteries using mechanical stress. *Journal of Power Sources*, 269:7–14, 2014.

[53] Y. Song, B. Lu, X. Ji, and J. Zhang. Diffusion Induced Stresses in Cylindrical Lithium-Ion Batteries: Analytical Solutions and Design Insights. *Journal of the Electrochemical Society*, 12(159):A2060–A2068, 2012.

[54] D. Chen. *Microscopic Investigations of Degradation in Lithium-Ion Batteries.* PhD thesis, KIT Karlsruhe, 2012.

[55] W.-S. Yoon, K. Y. Chung, J. McBreen, and X.-Q. Yang. A comparative study on structural change of $LiCo_{1/3}Ni_{1/3}Mn_{1/3}O_2$ and $LiNi_{0.8}Co_{0.15}Al_{0.05}O_2$ during first charge using in situ XRD. *Electrochemistry Communications*, 8:1257–1262, 2006.

[56] W.-J. Zhang. A review of electrochemical performance of alloy anodes for lithium-ion batteries. *Journal of Power Sources*, 196:13–24, 2011.

[57] D. Wang, X. Wu, Z. Wang, and L. Chen. Crack causing cyclic instability of $LiFePO_4$ cathode material. *Journal of Power Sources*, 140:125–128, 2004.

[58] S. Bhattacharya, A. R. Riahi, and A. T. Alpas. A transmission electron microscopy study of crack formation and propagation in electrochemically cycled graphite electrode in lithium-ion cells. *Journal of Power Sources*, 196:8719–8727, 2011.

[59] D. Chen, S. Indris, M. Schulz, B. Gamer, and R. Mönig. *In situ* electron microscopy on lithium-ion battery electrodes using an ionic liquid. *Journal of Power Sources*, 196:6382–6387, 2010.

[60] H. Gabrisch, J. Wilcox, and M. M. Doeff. TEM Study of Fracturing in Spherical and Plate-like $LiFePO_4$ Particles. *Electrochemical and Solid-State Letters*, 11(3):A25–A29, 2008.

[61] A. Chakraborty and N. Ramakrishnan. Prediction of electronic conductivity of a degrading electrode material using finite element method. *Computational Materials Science*, 69:455–465, 2013.

[62] T. G. Zavalis, M. Klett, M. H. Kjell, M. Behm, R. W. Lindström, and G. Lindbergh. Aging in lithium-ion batteries: Model and experimental investigation of harvested $LiFePO_4$ and a mesocarbon microbead graphite electrodes. *Electrochimica Acta*, 110:335–348, 2013.

[63] R. Deshpande, M. Verbrugge, Y.-T. Cheng, J. Wang, and P. Liu. Battery Cycle Life Prediction with Coupled Chemical Degradation and Fatigue Mechanics. *Journal of the Electrochemical Society*, 159(10):A1730–A1738, 2012.

[64] R. Narayanrao, M. M. Joglekar, and S. Inguva. A Phenomenological Degradation Model for Cyclic Aging of Lithium Ion Cell Materials. *Journal of the Electrochemical Society*, 160(1):A125–A137, 2013.

[65] E. Markevich, G. Salitra, M. D. Levi, and D. Aurbach. Capacity fading of lithiated graphite electrodes studied by a combination of electroanalytical methods, Raman spectroscopy and SEM. *Journal of Power Sources*, 146:146–150, 2005.

[66] J. Christensen and J. Newman. A Mathematical Model of Stress Generation and Fracture in Lithium Manganese Oxide. *Journal of the Electrochemical Society*, 153(6):A1019–A1030, 2006.

[67] X. Zhang, W. Shyy, and A. M. Sastry. Numerical Simulation of Intercalation-Induced Stress in Li-Ion Battery Electrode Particles. *Journal of the Electrochemical Society*, 154(10):A910–A916, 2007.

[68] Y.-T. Cheng and M. W. Verbrugge. Evolution of stress within a spherical insertion electrode particle under potentiostatic and

galvanostatic operation. *Journal of Power Sources*, 190:453–460, 2009.

[69] Y.-T. Cheng and M. W. Verbrugge. The influence of surface mechanics on diffusion induced stresses within spherical nanoparticles. *Journal of Applied Physics*, 104:083521, 2008.

[70] R. Purkayastha and R. M. McMeeking. A linearized model for Lithium Ion Batteries and Maps for their Performance and Failure. *Journal of Applied Mechanics*, 79(3):031021, 2012.

[71] B. Suthar and V. R. Subramanian. Lithium Intercalation in Core-Shell Materials–Theoretical Analysis. *Journal of the Electrochemical Society*, 161(5):A682–A692, 2014.

[72] F. Hao and D. Fang. Tailoring diffusion-induced stresses of core-shell nanotube electrodes in lithium-ion batteries. *Journal of Applied Physics*, 113:013507, 2013.

[73] K. Zhao, M. Pharr, L. Hartle, J. J. Vlassak, and Z. Suo. Fracture and debonding in lithium-ion batteries with electrodes of hollow core-shell nanostructures. *Journal of Power Sources*, 218:6–14, 2012.

[74] S. Y. Luchkin, H.-Y. Amanieu, D. Rosato, and A. L. Kholkin. Li distribution in graphite anodes: A Kelvin Probe Force Microscopy. *Journal of Power Sources*, 268:887–894, 2014.

[75] M.-K. Song, S. Park, F. M. Alamgir, J. Cho, and M. Liu. Nanostructured electrodes for lithium-ion and lithium-air batteries: the latest developments, challenges, and perspectives. *Material Science and Engineering R*, 72:203–252, 2011.

[76] Y. Yao, M. T. McDowell, I. Ryu, H. Wu, N. Liu, L. Hu, W. D. Nix, and Y. Cui. Interconnected Silicon Hollow Nanospheres for Lithium-Ion Battery Anodes with Long Cycle Life. *Nano Letters*, 11:2949–2954, 2011.

[77] M. Okubo, Y. Mizuno, H. Yamada, J. Kim, E. Hosono, H. Zhou, T. Kudo, and I. Honma. Fast Li-Insertion into Nanosized $LiMn_2O_4$ without Domain Boundaries. *ACS Nano*, 4(2):741–752, 2010.

[78] M. Wagemaker, F. M. Mulder, and A. Van der Ven. The Role of Surface and Interface Energy on Phase Stability of Nanosized Insertion Compounds. *Advanced Materials*, 21:2703–2709, 2009.

[79] A. Van der Ven and M. Wagemaker. Effect of surface energies and nano-particle size distribution on open circuit voltage of Li-electrodes. *Electrochemistry Communications*, 11:881–884, 2009.

[80] C. M. DeLuca, K. Maute, and M. L. Dunn. Effects of electrode particle morphology on stress generation on silicon during lithium insertion. *Journal of Power Sources*, 196:9672–9681, 2011.

[81] F. Hao, X. Gao, and D. Fang. Diffusion-induced stresses of electrode nanomaterials in lithium-ion battery: The effects of surface stress. *Journal of Applied Physics*, 112:103507, 2012.

[82] J.-L. Zang and Y.-P. Zhao. A diffusion and curvature dependent surface elastic model with application to stress analysis in lithium ion battery. *International Journal of Engineering Science*, 61:156–170, 2012.

[83] P. Bai, D. A. Cogswell, and M. Z. Bazant. Suppression of Phase Separation in $LiFePO_4$ Nanoparticles During Battery Discharge. *Nano Letters*, 11:4890–4896, 2011.

[84] D. A. Cogswell and M. Z. Bazant. Coherency strain and the kinetics of phase separation in $LiFePO_4$. *ACS Nano*, 6(3):2215–2225, 2012.

[85] S. Dargaville and T. W. Farrell. A comparison of mathematical models for phase-change in high-rate $LiFePO_4$ cathodes. *Electrochimica Acta*, 111:474–490, 2013.

[86] M. Huttin and M. Kamlah. Phase-field modeling of stress generation in electrode particles of lithium ion batteries. *Applied Physics Letters*, 101:133902, 2012.

[87] A.-C. Walk, M. Huttin, and M. Kamlah. Comparison of phase-field model for intercalation induced stresses in electrode particles of lithium ion batteries for small and finite deformation theory. *European Journal of Mechanics A/Solids*, 48:74–82, 2014.

[88] C. V. Di Leo, E. Rejovitzky, and L. Anand. A Cahn-Hilliard-type phase-field theory for species diffusion coupled with large elastic deformations: Application to phase-separating Li-ion electrode materials. *Journal of the Mechanics and Physics of Solids*, 70:1–29, 2014.

[89] K. Zhao, M. Pharr, J. J. Vlassak, and Z. Suo. Inelastic hosts as electrodes for high-capacity lithium-ion batteries. *Journal of Applied Physics*, 109:016110, 2011.

[90] K. Zhao, M. Pharr, S. Cai, J. J. Vlassak, and Z. Suo. Large plastic deformation in High-Capacity Lithium-Ion Batteries Caused by Charge and Discharge. *Journal of the American Ceramic Society*, 94:S226–S235, 2011.

[91] L. Anand. A Cahn-Hilliard-type theory for species diffusion coupled with large elastic-plastic deformations. *Journal of the Mechanics and Physics of Solids*, 60:1983–2002, 2012.

[92] A. F. Bower, P. R. Guduru, and V. A. Sethuraman. A finite strain model of stress, diffusion, plastic flow, and electrochemical reactions in a lithium-ion half-cell. *Journal of the Mechanics and Physics of Solids*, 59:804–828, 2010.

[93] Z. Cui, F. Gao, and J. Qu. A finite deformation stress-dependent chemical potential and ist applications to lithium ion batteries. *Journal of the Mechanics and Physics of Solids*, 60:1280–1295, 2012.

[94] J. Li, Q. Fang, F. Liu, and Y. Liu. Analytical modeling of dislocation effect on diffusion induced stress in a cylindrical lithium ion battery electrode. *Journal of Power Sources*, 272:121–127, 2014.

[95] W. H. Woodford, Y.-M. Chiang, and W. C. Carter. Design criteria for electrochemical shock resistant battery electrodes. *Energy & Environmental Science*, 5:8014, 2012.

[96] W. H. Woodford, Y.-M. Chiang, and W. C. Carter. Electrochemical Shock in Ion-Intercalation Materials with Limited Solid-Solubility. *Journal of the Electrochemical Society*, 160(8):A1286–A1292, 2013.

[97] K. Rhodes, M. Kirkham, R. Meisner, C. M. Parish, N. Dudney, and C. Daniel. Novel cell design for combined in situ acoustic emission and x-ray diffraction study during electrochemical cycling batteries. *Review of Scientific Instruments*, 82:075107, 2011.

[98] S. Kalnaus, K. Rhodes, and C. Daniel. A study of lithium intercalation induced fracture of silicon particles used as anode material in Li-ion battery. *Journal of Power Sources*, 196:8116–8124, 2011.

[99] T. Ohzuku, H. Tomura, and K. Sawai. Monitoring of Particle Fracture by Acoustic Emission during Charge and Discharge of Li/MnO_2 Cells. *Journal of the Electrochemical Society*, 10(144):3496–3500, 1997.

[100] N. Kircheva, S. Genies, C. Chabrol, and P.-X. Thivel. Evaluation of acoustic emission as a suitable tool for aging characterization of LiAl/$LiMnO_2$ cell. *Electrochimica Acta*, 88:488–494, 2013.

[101] F. Yang. Insertion-induced breakage of materials. *Journal of Applied Physics*, 108:073536, 2010.

[102] W. H. Woodford, Y.-M. Chiang, and W. C. Carter. Electrochemical shock of Intercalation Electrodes: A Fracture Mechanics Analysis. *Journal of the Electrochemical Society*, 157(10):A1052–A1059, 2010.

[103] M. Zhu, J. Park, and A. M. Sastry. Fracture Analysis of the Cathode in Li-Ion Batteries: A Simulation Study. *Journal of the Electrochemical Society*, 159(4):A492–A498, 2012.

[104] T. K. Bhandakkar and H. Gao. Cohesive modeling of crack nucleation in a cylindrical electrode under axisymmetric diffusion induced stresses. *International Journal of Solids and Structures*, 48:2304–2309, 2011.

[105] K. Zhao, M. Pharr, J. J. Vlassak, and Z. Suo. Fracture of electrodes in lithium-ion batteries caused by fast charging. *Journal of Applied Physics*, 108:073517, 2010.

[106] Y. Qi, Q. Xu, and A. Van der Ven. Chemically induced crack instability when electrodes fracture. *Journal of the Electrochemical Society*, 159(11):A1838–A1843, 2012.

[107] Y. F. Gao and M. Zhou. Coupled mechano-diffusional driving forces for fracture in electrode materials. *Journal of Power Sources*, 203:176–193, 2013.

[108] T. Rabczuk. Computational Methods for Fracture in Brittle and Quasi-Brittle Solids: State-of-the-Art Review and Future Perspectives. *ISRN Applied Mechanics*, 849231:1–38, 2013.

[109] C. Miehe, F. Welschinger, and M. Hofacker. Thermodynamically-consistent Phase Field Models of Fracture: Variational and Multi-Field FE Implementations. *International Journal for Numerical Methods in Engineering*, 83(10):1273–1311, 2010.

[110] M. J. Borden, T. J. R. Hughes, C. M. Landis, and C. V. Verhoosel. A higher-order phase-field model for brittle fracture: Formulation and analysis within the isogeometric analysis framework. *Computer Methods in Applied Mechanics and Engineering*, 273:100–118, 2014.

[111] R. de Borst, J. J. C. Remmers, A. Needleman, and M.-A. Abellan. Discrete vs smeared crack models for concrete fracture: bridging the gap. *International Journal for Numerical and Analytical Methods in Geomechanics*, 28(7-8):583–607, 2004.

[112] H. Amor, J.-J. Marigo, and C. Maurini. Regularized formulation of the variational brittle fracture with unliateral contact: Numerical experiments. *Journal of the Mechanics and Physics of Solids*, 57:1209–1229, 2009.

[113] A. R. Ingraffea and V. Saouma. *Numerical modeling of discrete crack propagation in reinforced plain concrete*, volume 4, pages 171–225. Fracture mechanics of concrete: Structural application and numerical calculation, 1985.

[114] N. Moës, J. Dolbow, and T. Belytschko. A finite element method for crack growth without remeshing. *International Journal for Numerical Methods in Engineering*, 46:131–150, 1999.

[115] M. Ortiz and A. Pandolfi. Finite-Deformation irreversible cohesive elements for three-dimensional crack-propagation analysis. *International Journal for Numerical Methods in Engineering*, 44:1267–1282, 1999.

[116] X.-P. Xu and A. Needleman. Numerical simulations of fast crack growth in brittle solids. *Journal of the Mechanics and Physics of Solids*, 42(9):1397–1434, 1994.

[117] M. J. Borden, C. V. Verhoosel, M. A. Scott, T. J. R. Hughes, and C. M. Landis. A phase-field description of dynamic brittle fracture. *Computer Methods in Applied Mechanics and Engineering*, 217-220:77–95, 2012.

[118] M. Hofacker and C. Miehe. Continuum phase field modeling of dynamic fracture: variational principles and staggered FE implementation. *International Journal of Fracture*, 178:113–129, 2012.

[119] M. Hofacker and C. Miehe. A phase field model of dynamic fracture: Robust field updates for the analysis of complex crack patterns. *International Journal for Numerical Methods in Engineering*, 93:276–301, 2013.

[120] C. Kuhn and R. Müller. Phase field simulation of thermomechanical fracture. *Proceedings in Applied Mathematics and Mechanics*, 9:191–192, 2009.

[121] B.-X. Xu, D. Schrade, D. Gross, and R. Müller. Fracture simulation of ferroelectrics based on the phase field continuum and a damage variable. *International Journal of Fracture*, 166:163–172, 2010.

[122] A. Abdollahi and I. Arias. Three-dimensional simulation of crack propagation in ferroelectric polycrystals: Effect of combined toughening mechanisms. *Acta Materialia*, 65:106–117, 2013.

[123] C. Miehe, F. Welschinger, and M. Hofacker. A phase field model of electromechanical fracture. *Journal of the Mechanics and Physics of Solids*, 58:1716–1740, 2010.

[124] Z. A. Wilson, M. J. Borden, and C. M. Landis. A phase-field model for fracture in piezoelectric ceramics. *International Journal of Fracture*, 183:135–153, 2013.

[125] D. Schneider, M. Selzer, J. Bette, I. Rementeria, A. Vondrous, M. J. Hoffmann, and B. Nestler. Phase-Field Modeling of Diffusion Coupled Crack Propagation Processes. *Advanced Engineering Materials*, 2(16):1527–2648, 2014.

[126] P. Zuo and Y.-P. Zhao. A phase field model coupling lithium diffusion and stress evolution with crack propagation and application in lithium ion. *Physical Chemistry Chemical Physics*, 2014.

[127] L. Liang, M. Stan, and M. Anitescu. Phase-field modeling of diffusion-induced crack propagations in electrochemical systems. *Applied Physics Letters*, 105:163903, 2014.

[128] Esther Bohn. *Partikel-Modell für Lithium-Diffusion und mechanische Spannungen einer Interkalationselektrode.* PhD thesis, KIT Karlsruhe, 2011.

[129] E. Bohn, T. Eckl, M. Kamlah, and R. M. McMeeking. A Model for Lithium Diffusion and Stress Generation in an Intercalation Storage Particle with Phase Change. *Journal of the Electrochemical Society*, 160(10):A1638–A1652, 2013.

[130] P. Haupt. *Continuum mechanics and theory of materials.* Springer Verlag, 2002.

[131] Y. F. Gao and M. Zhou. Strong stress-enhanced diffusion in amorphous lithium alloy nanowire electrodes. *Journal of Applied Physics*, 109:014310, 2011.

[132] T. L. Anderson. *Fracture Mechanics.* CRC Press, 1995.

[133] A. A. Griffith. The Phenomena of Rupture and Flow in Solids. *Philosophical Transactions, Series A*, 221:163–198, 1920.

[134] G. R. Irwin. Fracture Dynamics. *Fracturing of Metals*, 152, 1948.

[135] G. A. Francfort and J.-J. Marigo. Revisiting brittle fracture as an energy minimization problem. *Journal of the Mechanics and Physics of Solids*, 46(8):1319–1342, 1998.

[136] B. Bourdin, G. A. Francfort, and J.-J. Marigo. Numerical experiments in revisited brittle fracture. *Journal of the Mechanics and Physics of Solids*, 48:797–826, 2000.

[137] L. Ambrosio and V. M. Tortorelli. Approximation of Functionals Depending on Jumps by Elliptic Functionals via γ-Convergence. *Communications on Pure and Applied Mathematics*, 43:999–1036, 1990.

[138] D. Mumford and J. Shah. Optimal Approximations by Piecewise Smooth Functions and Associated Variational Problems. *Communications on Pure and Applied Mathematics*, 42:577–685, 1989.

[139] B. Bourdin, G. A. Francfort, and J.-J. Marigo. The variational approach to fracture. *Journal of Elasticity*, 91(1-3):5–148, 2008.

[140] C. Miehe, M. Hofacker, and F. Welschinger. A phase field model for rate-independent crack propagation: Robust algorithmic implementation based on operator splits. *Computer Methods in Applied Mechanics and Engineering*, 199:2765–2778, 2010.

[141] C. Kuhn and R. Müller. A continuum phase field model for fracture. *Engineering Fracture Mechanics*, 77:3625–3634, 2010.

[142] G. Del Piero, G. Lancioni, and R. March. A variational model for fracture mechanics: Numerical experiments. *Journal of the Mechanics and Physics of Solids*, 55:2513–2537, 2007.

[143] G. Lancioni and G. Royer-Carfagni. The Variational Approach to Fracture Mechanics. A Practical Application to the French Panthéon in Paris. *Journal of Elasticity*, 1-2(95):1–30, 2009.

[144] T. K. Bhandakkar and H. Gao. Cohesive modeling of crack nucleation under diffusion induced stresses in a thin strip: Implications on the critical size for flaw tolerant battery electrodes. *International Journal of Solids and Structures*, 47:1424–1434, 2010.

[145] H. Haftbaradaran and J. Qu. A path-independent integral for fracture of solids under combined electrochemical and mechanical loadings. *Journal of the Mechanics and Physics of Solids*, 71:1–14, 2014.

[146] W. E. Schiesser. *The Numerical Method of Lines*. Academic Press, 1991.

[147] O. C. Zienkiewicz, R. L. Taylor, and J. Z. Hu. *The finite element method - its basis and fundamentals*. Elsevier, 2005.

[148] D. Boffi, F. Brezzi, and M. Fortin. *Mixed Finite Element Methods and Applications*. Springer Verlag, 2013.

[149] S. C. Brenner and L. R. Scott. *The Mathematical Theory of Finite Element Methods*. Springer Verlag, 2007.

[150] G. A. Holzapfel. *Nonlinear Solid Mechanics: A Continuum Approach for Engineering.* Wiley, 2000.

[151] J. Bonet and R. D. Wood. *Nonlinear Continuum Mechanics For Finite Element Analysis.* Cambridge University Press, 1997.

[152] C. Kuhn and R. Müller. Exponential Finite Elements for a Phase Field Fracture Model. *Proceedings in Applied Mathematics and Mechanics*, 10:121–122, 2010.

[153] deal.II. `https://www.dealii.org/`. Accessed: 2015-18-01.

[154] W. Bangerth, R. Hartmann, and G. Kanschat. deal.II - a General Purpose Object Oriented Finite Element Library. *ACM Transactions on Mathematical Software*, 33(4):24, 2007.

[155] W. Bangerth, C. Burstedde, T. Heister, and M. Kronbichler. Algorithms and Data Structures fo Massively Parallel Generic Adaptive Finite Element Codes. *ACM Transactions on Mathematical Software*, 38(14):1–28, 2011.

[156] M. A. Heroux, R. A. Bartlett, V. E. Howle, R. J. Hoekstra, J. J. Hu, T. G. Kolda, R. B. Lehoucq, K. R. Long, R. P. Pawlowski, E. T. Phipps, A. G. Salinger, H. K. Thornquist, R. S. Tuminaro, J. M. Willenbring, A. Williams, and K. S. Stanley. An overview of the Trilinos project. *ACM Transactions on Mathematical Software*, 31(3):397–423, 2005.

[157] S. Balay, W. D. Gropp, L. Curfman McInnes, and B. F. Smith. *Modern Software Tools in Scientific Computing - Efficient Management of Parallelism in Object Oriented Numerical Software Libraries.* Birkhäuser Press, 1997.

[158] ParaView. `http://www.paraview.org/`. Accessed: 2014-11-02.

[159] Trelis. `http://www.csimsoft.com/trelis.jsp/`. Accessed: 2014-11-02.

[160] J. Nocedal and S. Wright. *Numerical Optimization (Springer Series in Operations Research and Financial Engineering).* Springer Verlag, 2006.

[161] H. Tada, P. C. Paris, and G. R. Irwin. *The Stress Analysis of Cracks Handbook.* Professional Engineering Publishing, 2000.

[162] J. Lemaitre. *A course on damage theory.* Springer Verlag, 1996.

[163] B. Bourdin. Numerical implementation of the variational formulation for quasi-static brittle fracture. *Interfaces and Free Boundaries*, 9:411–430, 2007.

[164] K. Pham, H. Amor, J.-J. Marigo, and C. Maurini. Gradient Damage Models and Their Use to Approximate Brittle Fracture. *International Journal of Damage Mechanics*, 20:618–652, 2011.

[165] K. Pham and J.-J. Marigo. Construction and analysis of localized responses for gradient damage models in a 1D setting. *Vietnam Journal of Mechanics*, 31(3&4):233–246, 2009.

[166] B. Budiansky and R. J. O'Connell. Elastic moduli of a cracked solid. *International Journal of Solids and Structures*, 12:81–97, 1976.

[167] R. Purkayastha and R. M. McMeeking. A parameter study of intercalation of lithium into storage particles in a lithium-ion battery. *Computational Materials Science*, 80:2–14, 2013.

[168] H.-Y. Amanieu, D. Rosato, M. Sebastiani, F. Massimi, and D. C. Lupascu. Mechanical property measurements of heterogenous materials by selective nanoindentation: Application to $LiMn_2O_4$ cathode. *Materials Science & Engineering A*, 593:92–102, 2014.

[169] H. Y. Amanieu. Nanomechanics of Battery composites: from nanoindentation to AFM. In *MSE 2014, Symposia D06*, 2014.

[170] R. E. Garcia, Y.-M. Chiang, W. C. Carter, P. Limthongkul, and C. M. Bishop. Microstructural Modeling and Design of Rechargeable Lithium-Ion Batteries. *Journal of the Electrochemical Society*, 152(1):A255–A263, 2005.

[171] S. Timoshenko and J. N. Goodier. *Theory of Elasticity.* Mcgraw-Hill Professional, 1970.

[172] M. Rahman and T. Michelitsch. A note on the formula for the Rayleigh wave speed. *Wave Motion*, 43:272–276, 2006.

[173] E. Katzav, M. Adda-Bedia, and R. Arias. Theory of dynamic crack branching in brittle materials. *International Journal of Fracture*, 143:245–271, 2007.

[174] M. Adda-Bedia. Brittle fracture dynamics with arbitrary paths III. The branching instability under general loading. *Journal of the Mechanics and Physics of Solids*, 53:227–248, 2004.

[175] L. B. Freund. *Dynamic fracture mechanics.* Cambridge University Press, 1990.

[176] K. B. Broberg. *Cracks and fracture.* Academic Press, 1999.

[177] R. V. Goldstein and R. L. Salganik. Brittle fracture of solids with arbitrary cracks. *International Journal of Fracture*, 10(4):507–523, 1974.

[178] J. B. Leblond. Crack paths in plane situations-I. General form of the expansion of the stress intensity factors. *International Journal of Solids and Structures*, 25:1311–1325, 1989.

[179] M. Adda-Bedia, R. Arias, M. Ben Amar, and F. Lund. Generalized Griffith criterion for dynamic fracture and the stability of crack motion at high velocities. *Physical Review E*, 60(2):2366–2376, 1999.

[180] B. V. Kostrov. On the crack propagation with variable velocity. *International Journal of Fracture*, 11(1):47–56, 1973.

[181] S. Nemat-Nasser, L. M. Keer, and K. S. Parihar. Unstable Growth of Thermally Induced Interacting Cracks in Brittle Solids. *International Journal of Solids and Structures*, 14:409–430, 1978.

[182] H.-A. Bahr, H. Balke, M. Kuna, and H. Liesk. Fracture Analysis of a Single Edge Cracked Strip Under Thermal Shock. *Theoretical and Applied Fracture Mechanics*, 8:33–39, 1987.

[183] P. Heitjans and J. Kärger. *Diffusion in Condensed Matter.* Springer Verlag, 2005.

Schriftenreihe
des Instituts für Angewandte Materialien

ISSN 2192-9963

Die Bände sind unter www.ksp.kit.edu als PDF frei verfügbar
oder als Druckausgabe bestellbar.

Band 1 Prachai Norajitra
Divertor Development for a Future Fusion Power Plant. 2011
ISBN 978-3-86644-738-7

Band 2 Jürgen Prokop
Entwicklung von Spritzgießsonderverfahren zur Herstellung von Mikrobauteilen durch galvanische Replikation. 2011
ISBN 978-3-86644-755-4

Band 3 Theo Fett
New contributions to R-curves and bridging stresses – Applications of weight functions. 2012
ISBN 978-3-86644-836-0

Band 4 Jérôme Acker
Einfluss des Alkali/Niob-Verhältnisses und der Kupferdotierung auf das Sinterverhalten, die Strukturbildung und die Mikrostruktur von bleifreier Piezokeramik $(K_{0,5}Na_{0,5})NbO_3$. 2012
ISBN 978-3-86644-867-4

Band 5 Holger Schwaab
Nichtlineare Modellierung von Ferroelektrika unter Berücksichtigung der elektrischen Leitfähigkeit. 2012
ISBN 978-3-86644-869-8

Band 6 Christian Dethloff
Modeling of Helium Bubble Nucleation and Growth in Neutron Irradiated RAFM Steels. 2012
ISBN 978-3-86644-901-5

Band 7 Jens Reiser
Duktilisierung von Wolfram. Synthese, Analyse und Charakterisierung von Wolframlaminaten aus Wolframfolie. 2012
ISBN 978-3-86644-902-2

Band 8 Andreas Sedlmayr
Experimental Investigations of Deformation Pathways in Nanowires. 2012
ISBN 978-3-86644-905-3

Band 9 Matthias Friedrich Funk
Microstructural stability of nanostructured fcc metals during cyclic deformation and fatigue. 2012
ISBN 978-3-86644-918-3

Band 10 Maximilian Schwenk
Entwicklung und Validierung eines numerischen Simulationsmodells zur Beschreibung der induktiven Ein- und Zweifrequenzrandschichthärtung am Beispiel von vergütetem 42CrMo4. 2012
ISBN 978-3-86644-929-9

Band 11 Matthias Merzkirch
Verformungs- und Schädigungsverhalten der verbundstranggepressten, federstahldrahtverstärkten Aluminiumlegierung EN AW-6082. 2012
ISBN 978-3-86644-933-6

Band 12 Thilo Hammers
Wärmebehandlung und Recken von verbundstranggepressten Luftfahrtprofilen. 2013
ISBN 978-3-86644-947-3

Band 13 Jochen Lohmiller
Investigation of deformation mechanisms in nanocrystalline metals and alloys by in situ synchrotron X-ray diffraction. 2013
ISBN 978-3-86644-962-6

Band 14 Simone Schreijäg
Microstructure and Mechanical Behavior of Deep Drawing DC04 Steel at Different Length Scales. 2013
ISBN 978-3-86644-967-1

Band 15 Zhiming Chen
Modelling the plastic deformation of iron. 2013
ISBN 978-3-86644-968-8

Band 16 Abdullah Fatih Çetinel
Oberflächendefektausheilung und Festigkeitssteigerung von niederdruckspritzgegossenen Mikrobiegebalken aus Zirkoniumdioxid. 2013
ISBN 978-3-86644-976-3

Band 17 Thomas Weber
Entwicklung und Optimierung von gradierten Wolfram/ EUROFER97-Verbindungen für Divertorkomponenten. 2013
ISBN 978-3-86644-993-0

Band 18 Melanie Senn
Optimale Prozessführung mit merkmalsbasierter Zustandsverfolgung. 2013
ISBN 978-3-7315-0004-9

Band 19 Christian Mennerich
Phase-field modeling of multi-domain evolution in ferromagnetic shape memory alloys and of polycrystalline thin film growth. 2013
ISBN 978-3-7315-0009-4

Band 20 Spyridon Korres
On-Line Topographic Measurements of Lubricated Metallic Sliding Surfaces. 2013
ISBN 978-3-7315-0017-9

Band 21 Abhik Narayan Choudhury
Quantitative phase-field model for phase transformations in multi-component alloys. 2013
ISBN 978-3-7315-0020-9

Band 22 Oliver Ulrich
Isothermes und thermisch-mechanisches Ermüdungsverhalten von Verbundwerkstoffen mit Durchdringungsgefüge (Preform-MMCs). 2013
ISBN 978-3-7315-0024-7

Band 23 Sofie Burger
High Cycle Fatigue of Al and Cu Thin Films by a Novel High-Throughput Method. 2013
ISBN 978-3-7315-0025-4

Band 24 Michael Teutsch
Entwicklung von elektrochemisch abgeschiedenem LIGA-Ni-Al für Hochtemperatur-MEMS-Anwendungen. 2013
ISBN 978-3-7315-0026-1

Band 25 Wolfgang Rheinheimer
Zur Grenzflächenanisotropie von $SrTiO_3$. 2013
ISBN 978-3-7315-0027-8

Band 26 Ying Chen
Deformation Behavior of Thin Metallic Wires under Tensile and Torsional Loadings. 2013
ISBN 978-3-7315-0049-0

Band 27 Sascha Haller
Gestaltfindung: Untersuchungen zur Kraftkegelmethode. 2013
ISBN 978-3-7315-0050-6

Band 28 Stefan Dietrich
Mechanisches Verhalten von GFK-PUR-Sandwichstrukturen unter quasistatischer und dynamischer Beanspruchung. 2013
ISBN 978-3-7315-0074-2

Band 29 Gunnar Picht
Einfluss der Korngröße auf ferroelektrische Eigenschaften dotierter $Pb(Zr_{1-x}Ti_x)O_3$ Materialien. 2013
ISBN 978-3-7315-0106-0

Band 30 Esther Held
Eigenspannungsanalyse an Schichtverbunden mittels inkrementeller Bohrlochmethode. 2013
ISBN 978-3-7315-0127-5

Band 31 Pei He
On the structure-property correlation and the evolution of Nanofeatures in 12-13.5% Cr oxide dispersion strengthened ferritic steels. 2014
ISBN 978-3-7315-0141-1

Band 32 Jan Hoffmann
Ferritische ODS-Stähle – Herstellung, Umformung und Strukturanalyse. 2014
ISBN 978-3-7315-0157-2

Band 33 Wiebke Sittel
Entwicklung und Optimierung des Diffusionsschweißens von ODS Legierungen. 2014
ISBN 978-3-7315-0182-4

Band 34 Osama Khalil
Isothermes Kurzzeitermüdungsverhalten der hoch-warmfesten Aluminium-Knetlegierung 2618A (AlCu2Mg1,5Ni). 2014
ISBN 978-3-7315-0208-1

Band 35 Magalie Huttin
Phase-field modeling of the influence of mechanical stresses on charging and discharging processes in lithium ion batteries. 2014
ISBN 978-3-7315-0213-5

Band 36 Christoph Hage
Grundlegende Aspekte des 2K-Metallpulverspritzgießens. 2014
ISBN 978-3-7315-0217-3

Band 37 Bartłomiej Albiński
Instrumentierte Eindringprüfung bei Hochtemperatur für die Charakterisierung bestrahlter Materialien. 2014
ISBN 978-3-7315-0221-0

Band 38 Tim Feser
Untersuchungen zum Einlaufverhalten binärer alpha-Messinglegierungen unter Ölschmierung in Abhängigkeit des Zinkgehaltes. 2014
ISBN 978-3-7315-0224-1

Band 39 Jörg Ettrich
Fluid Flow and Heat Transfer in Cellular Solids. 2014
ISBN 978-3-7315-0241-8

Band 40 Melanie Syha
Microstructure evolution in strontium titanate Investigated by means of grain growth simulations and x-ray diffraction contrast tomography experiments. 2014
ISBN 978-3-7315-0242-5

Band 41 Thomas Haas
Mechanische Zuverlässigkeit von gedruckten und gasförmig abgeschiedenen Schichten auf flexiblem Substrat. 2014
ISBN 978-3-7315-0250-0

Band 42 Aron Kneer
Numerische Untersuchung des Wärmeübertragungsverhaltens in unterschiedlichen porösen Medien. 2014
ISBN 978-3-7315-0252-4

Band 43 Manuel Feuchter
Investigations on Joule heating applications by multiphysical continuum simulations in nanoscale systems. 2014
ISBN 978-3-7315-0261-6

Band 44 Alexander Vondrous
Grain growth behavior and efficient large scale simulations of recrystallization with the phase-field method. 2014
ISBN 978-3-7315-0280-7

Band 45 Tobias Kennerknecht
Fatigue of Micro Molded Materials – Aluminum Bronze and Yttria Stabilized Zirconia. 2014
ISBN 978-3-7315-0293-7

Band 46 Christopher Scherr
Elektrochemisches Verhalten von Lithium-Schwefel-Zellen mit unterschiedlicher Kathodenstruktur. 2015
ISBN 978-3-7315-0296-8

Band 47 Konstantin Frölich
Der Decal-Prozess zur Herstellung katalysatorbeschichteter Membranen für PEM-Brennstoffzellen. 2015
ISBN 978-3-7315-0334-7

Band 48 Benedikt Haspel
Werkstoffanalytische Betrachtung der Eigenschaften von mittels neuartiger RTM-Fertigungsprozesse hergestellten glasfaserverstärkten Polymerverbunden. 2015
ISBN 978-3-7315-0337-8

Band 49 Marco Berghoff
Skalenübergreifende Modellierung und Optimierung vom atomistischen kristallinen Phasenfeldmodell bis zur mesoskopischen Phasenfeldmethode. 2015
ISBN 978-3-7315-0416-0

Band 50 Michael Selzer
Mechanische und Strömungsmechanische Topologieoptimierung mit der Phasenfeldmethode. 2016
ISBN 978-3-7315-0431-3

Band 51 Michael Mahler
Entwicklung einer Auswertemethode für bruchmechanische Versuche an kleinen Proben auf der Basis eines Kohäsivzonenmodells. 2016
ISBN 978-3-7315-0441-2

Band 52 Christoph Bohnert
Numerische Untersuchung des Verformungs- und Bruchverhaltens von einkristallinem Wolfram auf mikroskopischer Ebene. 2016
ISBN 978-3-7315-0444-3

Band 53 Stefan Guth
Schädigung und Lebensdauer von Nickelbasislegierungen unter thermisch-mechanischer Ermüdungsbeanspruchung bei verschiedenen Phasenlagen. 2016
ISBN 978-3-7315-0445-0

Band 54 Markus Klinsmann
The Effects of Internal Stress and Lithium Transport on Fracture in Storage Materials in Lithium-Ion Batteries. 2016
ISBN 978-3-7315-0455-9